SHAPING THINGS

by

BRUCE STERLING

DESIGNER
LORRAINE WILD

EDITORIAL DIRECTOR
PETER LUNENFELD

MEDIAWORK
The MIT Press
Cambridge, Massachusetts
London, England

CONTENTS

1.

TO WHOM
IT OUGHT TO CONCERN

This book is about created objects and the environment, which is to say, it's a book about everything. Seen from sufficient distance, this is a small topic.

The ideal readers for this book are those ambitious young souls (of any age) who want to constructively intervene in the process of technosocial transformation. That is to say, this book is for designers and thinkers, engineers and scientists, entrepreneurs and financiers, and anyone else who might care to understand why things were once as they were, why things are as they are, and what things seem to be becoming.

The world of organized artifice is transforming in ways that are poorly understood and little explored. There are two reasons why this is happening.

First, new forms of design and manufacture are appearing that lack historical precedent, and are bound to create substantial novelty.

Second, the production methods currently used are not sustainable. They are large in scale, have long histories,

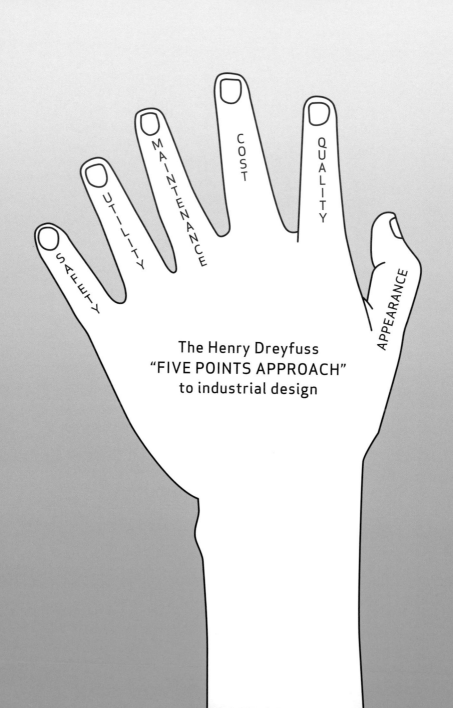

SAFETY

UTILITY

MAINTENANCE

COST

QUALITY

APPEARANCE

The Henry Dreyfuss
"FIVE POINTS APPROACH"
to industrial design

and have been extensively researched and developed, but they can't go on in their present form. The status quo uses archaic forms of energy and materials which are finite and toxic. They wreck the climate, poison the populace and foment resource wars. They have no future.

So the challenge at hand is to creatively guide the tremendous vectors of the first reason, so as to finesse the horrific consequences of the second reason. Then we can enjoy some futurity. That's what this book is about.

The quest for a sustainable world may succeed, or it may fail. If it fails, the world will become unthinkable. If it works, the world will become unimaginable.

In practice, people will experience mixed success. So tomorrow's world will be partially unthinkable and partially unimaginable. Effective actors will be poised between these two conditions, with plans and crowbars.

I hope this book will be a stimulating contribution to the blueprinting effort.

I hope you enjoy reading this half as much as I enjoyed speculating about it.

FIGURE 1

2.

TOMORROW COMPOSTS TODAY

I'm using idiosyncratic terms that might become confusing outside this context of discussion. So, I'm going to CAPITALIZE them: "*Artifacts*," "**MACHINES**," "*PRODUCTS*," "GIZMOS." "SPIME" is a flat-out neologism, but central to the thesis of this book, so I'll CAPITALIZE that too. This will emphasize that I'm talking about *classes* of objects in varying object-human relationships, rather than some particular *Artifact*, **MACHINE**, *PRODUCT*, GIZMO or SPIME.

By using this special terminology, I want to emphasize the continuing interplay between objects and people. I'm describing an infrastructure of human support, irrevocably bound to and generated by the class of people who are necessary to create and maintain that infrastructure. It's mentally easier to divide humans and objects than to understand them as a comprehensive and interdependent system: people are alive, objects are inert, people can think, objects just lie there. But this taxonomical division blinds us to the ways and means by which objects do change, and it obscures the areas of intervention where design can reshape things. Effective intervention takes place not in the human, not in the object, but in the realm of the technosocial.

So, by capital-*a* "*Artifacts*," I mean simple artificial objects, made by hand, used by hand, and powered by muscle. *Artifacts* are created one at time, locally, by rules of thumb and folklore rather than through any abstract understanding of the principles of mechanics. People within an infrastructure of *Artifacts* are "Hunters and Farmers."

By "**MACHINES**" I mean complex, precisely propor-tioned artifacts with many integral moving parts that have tapped some non-human, non-animal power source. **MACHINES** require specialized support structures for engineering skills, distribution, and finance. People within an infrastructure of **MACHINES** are "Customers."

So what's the difference?

How does one draw the line between a technoculture of *Artifacts* and a technoculture of **MACHINES**?

I draw two lines of division. The first line is the Line of No Return. The second is the Line of Empire.

We know there has been a revolution in technoculture when that technoculture cannot voluntarily return to the previous technocultural condition. A sailor can become a farmer, but if the sailors from the **MACHINE** era of iron and steam return to the earlier *Artifact* era of wood and sail, millions will starve to death. The technosociety will collapse, so it's no longer an option. That's the <u>Line of No Return</u>.

We know that this revolution has become the new status quo when even the fiercest proponents of the earlier technoculture cannot physically overwhelm and defeat the new one.

The new technoculture's physical advantages in shaping objects make it impregnable. The imperial technoculture can spew its objects and processes abroad, more or less at will.

Those who lack that productive capacity are forced into colonial or defensive postures. That's the <u>Line of Empire</u>.

I'm therefore inclined to date the advent of **MACHINE** technoculture to the eclipse of the Mongols in the 1500s. Before that time, an *Artifact* culture with bows and horses could blacken the earth with its rampaging hordes. After that date, the world is at the mercy of the West, as mechanization takes command.

By "*PRODUCTS*" I mean widely distributed, commercially available objects, anonymously and uniformly manufactured in massive quantities, using a planned division of labor, rapid, non-artisanal, assembly-line techniques, operating over continental economies of scale, and supported by highly reliable transportation, finance and information systems. People within an infrastructure of *PRODUCTS* are "<u>Consumers.</u>"

I would date the advent of *PRODUCT* technoculture to the period around World War One.

"GIZMOS" are highly unstable, user-alterable, baroquely multifeatured objects, commonly programmable, with a brief lifespan. GIZMOS offer functionality so plentiful that it is cheaper to import features into the object than it is to simplify it. GIZMOS are commonly linked to network service providers; they are not stand-alone objects but interfaces. People within an infrastructure of GIZMOS are "End-Users."

Unlike *Artifacts*, **MACHINES**, and *PRODUCTS*, GIZMOS have enough functionality to actively nag people. Their deployment demands extensive, sustained interaction: upgrades, grooming, plug-ins, plug-outs, unsought messages, security threats, and so forth.

The GIZMO epoch begins in 1989.

"SPIMES" are manufactured objects whose informational support is so overwhelmingly extensive and rich that they are regarded as material instantiations of an immaterial system. SPIMES begin and end as data. They are designed on screens, fabricated by digital means, and precisely tracked through space and time throughout their earthly sojourn.

SPIMES are sustainable, enhanceable, uniquely identifiable, and made of substances that can and will be folded back into the production stream of future SPIMES. Eminently data-mineable, SPIMES are the protagonists of an historical process.

People within an infrastructure of SPIMES are "Wranglers."

I would date the dawn of SPIMES to 2004, when the United States Department of Defense suddenly demanded that its thousands of suppliers attach Radio Frequency ID tags, or "arphids," to military supplies. If this innovation turns out to be of genuine military advantage, and if it also spreads widely in commercial inventory systems, then a major transition will likely be at hand.

SPIMES are coming sooner or later, for SPIMES are here in primitive forms already. We can't yet know if this is an important development, or just a visionary notion. The technical potential seems quite large, but how much design energy will these opportunities attract? Who will dare to use these potentials as a means of technosocial intervention? Is there a Line of No Return, and a Line of Empire? And if so, where are those lines?

When will we realize that we need these structures in order to live—that we can't surrender their advantages without awful consequence? And when will polities infested with SPIMES realize that they can lord it over those who refuse or fail to adapt them?

If I had to guess, I'd say 30 years. In 30 years, things properly understood as SPIMES will be all around us. Mind you, this is by no means an entirely happy prospect. It's important to explicitly acknowledge the downsides of any technological transformation—to "think of the underside first," to think in a precautionary way. In engaging with a technology so entirely friendly toward surveillance, spying, privacy invasion and ruthless technical intrusion on previously unsoiled social spaces, we are playing with fire. Nothing new there—fire is two million years old. It helps to learn about fire and its remarkable affordances. Not a lot is to be gained by simply flinging lit matches.

Design thinking and design action should be the proper antidotes to fatalistic handwringing when it comes to technology's grim externalities and potentials for deliberate abuse. This book is for designers who want to be active agents in a technosocial world. I can't make you into a moral angel (because I'm not one myself and have little interest in being one), but I might help you understand that the future can be yours to make.

Of course that's not the end of the story. The story, if it's successful, fails to end because we have created SPIMES and can manage them successfully. By handling challenges properly, we've enjoyed life without spoiling it for our descendants; as a culture, we've obtained more future. That would be the victory condition and the point of writing books of this kind.

I'll be spending most of the rest of this little book exploring what a SPIME might be, or become, and how people will interact with SPIMES. There are no such things as true SPIMES yet—these are still speculative, imaginary concepts. I will try to make the case that SPIMES are genuine prospects for genuine objects in the future, and worthy of designers' attention. I hope to persuade you that clever young people had better get used to these ideas.

Tomorrow composts today.

In other words, technocultures do not abolish one another in clean or comprehensive ways. Instead, new capacities are layered onto older ones. The older technosocial order gradually loses its clarity, crumbles, and melts away under the accumulating weight of the new.

The coming advent of SPIMES will not "abolish" the dominant technoculture we see today, which is the GIZMO. *Artifacts*, PRODUCTS and **MACHINES** are still plentiful and flourish in today's GIZMO world—but, influenced by the pressure from on high, they do tend to take on a pervasive flavor of GIZMO. Let's see how.

3.
OLD WINE
IN NEW BOTTLES

With a parable of old wine in new bottles, I can illustrate that some objects are born GIZMO, while others have GIZMOlogy thrust upon them. Let's get very immediate, practical and hands-on with the topic: let me invite you to sit down and have a glass of wine with me.

This wine bottle we have at hand here would seem to be a pretty simple object. It is mostly made of a single, ancient substance—glass—and it has no moving parts (if you don't count the discarded cork). Wine bottles are ancient, pre-industrial, even pre-historic. When I turn over this wine bottle, pour it and drink the contents, I'm experiencing the same somatic shock as Socrates.

But Socrates (who was a Hunter-Farmer from a world of *Artifacts*) was drinking local wine from a Greek vineyard in a handmade clay krater. Whereas I am an End-User in a technosociety dominated by GIZMOS. So I am drinking from a machine-labeled, mass-produced bottle of industrial glass, with a barcode and legalistic health warnings, which exists in many hundreds of identical copies, and was shipped from Italy to California and offered for sale in a vast supermarket.

And yes, this bottle of wine has a Webpage. This is how it leans forward into the future world of the SPIME.

SANGIOVESE
(san-joh-VEH-seh)
"From the Emilia-Romagna area comes Sangiovese, an easy to drink dry wine with a straightforward, spicy-fruity flavor. Enjoy this wine with a wide variety of foods including pasta with light tomato sauce, ribs, chicken, veal, pork, beef, cold meats and cold salads."
Winemaker notes (requires Adobe Acrobat Reader)
Here's how to say it
Host a tasting
Classics
Soave | Valpolicella | Bardolino
Varietals
Pinot Grigio | Merlot | Sangiovese | Chardonnay |
Cabernet Sauvignon
Signature Series
Arcale | Le Poiane | Tufaie | Colforte
Alta Gamma
Creso | Amarone

I had to use a laptop computer to access that data, but since I am an End-User of GIZMOS, I am rarely without a laptop computer. That was an affectation earlier in my lifespan, but I have crossed the Line of No Return with my laptop GIZMO here; I can no longer earn a living without it.

So this Sangiovese may be a "classic" wine from the Mediterranean basin, but this bottle is no longer a classic Artifact. It is GIZMO-ized.

Consider the wide variety of ways in which I am being invited to interact with this wine bottle. I don't merely drink the contents. I *could* just drink it—but if I lift my eyes just a little—(it took me 35 seconds, on wireless broadband, using the very machine on which I am writing this book)—then I am invited to learn how to pronounce a foreign language, how to set up a social gathering with my friends, how the wine is made (that might require me to download some software, mind you), and how to expand my oenophilic knowledge of grape varieties.

This is *no accident*. There is nothing frivolous or extraneous about this sudden explosion of informational intimacy between myself and a bottle of wine.

Every one of these transitions—Artifact to **MACHINE** to PRODUCT to GIZMO—involves an expansion of information. It enables a deeper, more intimate, more multiplex interaction between humans and objects.

In an Artifact technoculture, literacy is a frill. Scribes are hard to train, few in number and expensive to support.

The Hunter and Farmer lives close to the soil in a life bound to the rhythm of the seasons.

In a **MACHINE** technoculture, I am a Customer. I have a medium of exchange that commands a gamut of objects, plus banks, roads, ships, commercial records, engineering manuals, treatises on architecture, and a host of technical specialists engaged in craft. I'm literate and numerate, for the lack of such informational skills would put me at severe disadvantage.

In a PRODUCT technoculture, I'm a Consumer. Goods are available at commodity prices in a literally unknowable profusion. PRODUCTS are so radiantly specialized that they can be aimed with precision at defined consumer demographics: high-end, mid-list, down-market. I'm barked at by incessant advertising—unsought information flows—and burdened with mail-order catalogs. PRODUCTS may dare to have Some Assembly Required, but if so, I'll complain about that—for I am a Consumer, and want to be catered to. I exist under pressure of catering. To be catered to is my very life: I'm a social security number, a driver's license, a voter registration card, a stock portfolio and a retirement plan. Withdraw those structures, and I don't live.

In a GIZMO technoculture, my products are festooned with baroque amounts of functionality and tied deeply into sophisticated, unstable networks of service provision. As an End-User in a destabilized high tech society, I take great comfort in useless functions; they may well be impractical, but they give me a sense of dignity, like the silk ribbons and gold braid on a Renaissance courtier.

A device that is simple and easy to understand is a mere commodity; in my GIZMO society, mere MACHINES and PRODUCTS offer a poor return on investment. These crude devices lack the dams and weirs and tidal pools of patents and intellectual property; they offer no arena for bravura displays of my technical mastery. I rather prefer my devices not to work quite properly. I am balanced on the edge of complexity and utter chaos.

GIZMO wine hasn't ceased to be wine. Wine fanciers can take comfort in wine's long and successful historical record. However, this bottle with the Web site on its label isn't *Artifact* wine. The grapes were cultivated with fossil-fueled tractors, but this isn't MACHINE wine, either. It was shipped across the planet, tax-stamped and offered for sale in a supermarket, but it isn't PRODUCT wine.

This is GIZMO wine. It is offering me more functionality than I will ever be able to explore. This wine bottle aims to *educate me*—it is luring me to become more knowledgeable about the people and processes that made the bottle and its contents. It wants me to recruit me to become an unpaid promotional agent, a wine critic, an opinion maker—it wants me to throw wine-tasting parties and tell all my friends about my purchase. It is acculturating me to GIZMO technosociety.

Like all GIZMOS, its lifespan is brief—it doesn't take long to drink a bottle of wine. An individual bottle of wine, my entree into this wondrously elaborate process, costs a modest sum. The cost of the information jamboree that accompanies it has been amortized across a huge global base of willing consumers of flavored alcohol.

This GIZMO aspect of the wine bottle, all this Web page busy-ness and the bar code and the health warnings—these are not fripperies. They were all designed with deliberate care. Some were legislated. They are permanent changes in the relation between humans and wine bottles. Wine merchants will not retreat from this new digitized complexity with the purchaser because they have already installed that complexity throughout the rest of their production system and supply chain: from Italian agribusiness, through European oversight and standards, to distribution centers and retail outlets—none of them run blind any more, they are all linked through electronic commerce.

So why should I, the buyer, be left out? I *can't* be left out. Every producer and seller of GIZMOS is a buyer of somebody else's GIZMOS—the older roles of buyer, seller, producer, developer are all melted down in the informational stew. It costs very little to drag me into the digital mire.

What does this mean in practice—"dragged into the digital mire"?

It means taking my money of course, but money is often a metric proxy for two other, vaguer phenomena: cognitive load and opportunity costs.

To participate in the GIZMOS world, I need to think about things, talk about things, pay attention to things, be entertained by things... I pay a price for that in personal brainpower. That price is my own cognitive load. In a GIZMO world, I can learn a great deal about wine if I like, and that may even be cunningly arranged for me as a seductive, congenial, infotainment process—but if I do that, then I'll have to think less, or more hastily, or more sketchily, about some other things.

So I'll have to choose options, or at least navigate the risks. For instance: will I stop to read all the shrink-wrapped complexities of my software's End-User License Agreement? Or will I just hastily click AGREE and hope I'm too small-time to get sued?

Along with thinking less comes doing less: "opportunity cost". To make room in my life for this GIZMO jamboree, I have to sacrifice something that I'm already doing. There are only so many hours in the day, so there will be something I can't and don't do much any more. I pay a price in opportunity —maybe just the opportunity to sit still, like Socrates, unperturbed, in an olive grove, with my own unperturbed thoughts.

When it comes to objects, designers tend to be lavishly generous with their own cognitive load and opportunity cost. Thinking about objects is a designer's profession and avocation, and the chance to do more of that that is considered professional opportunity. But for people who are already fully booked mentally—the vast majority of the human race who aren't designers—those demands can be crippling.

> Everyone can't be a designer

—any more than everyone can be a mayor or a Senator.

There's not enough time in the world for people to sacrifice infinite amounts of opportunity and cognition. This means that, in a SPIME world, designers must design, not just for objects or for people, but for the technosocial interactions that unite people and objects: designing for opportunity costs and cognitive load. These resources deserve special design attention because these are the resources that are now in scarcity.

In a world of SPIME, the growing problems of attention load and opportunity costs have been finessed. Most probably, they've been deputized to powerful information machinery. These processes depend, as search engines do, on social software which can track human desire and interest.

What's basically missing in the future transition from GIZMO to SPIME are new, inventive, interactive machineries of representative design. As with representative government, these would be transparent and accountable infrastructures that could drag and Wrangle me into

the hurly-burly world of design issues without also crush-
ing me under the load of micro-management.

Can this be done? I think it can, if designers make it
happen. If done properly, it will be almost beneath notice.
People always do useful, supportive work for a technosocial system, whether they want to or not, whether they
know it or not. Hopefully, they can do it without the loss
of every precious instant in their life spans.

We interact with infrastructure differently in a world
with representative design. In particular, with enough
informational power, the "invisible hand of the market"
becomes visible. The hand of the market was called
"invisible" because Adam Smith, an eighteenth-century
economist, had very few ways to measure it. Adam Smith
lacked metrics. Metrics make things visible. In a SPIME
technosociety, most everything has metrics. Human
beings and their objects are awash in metrics. There are
many ways to make these metrics impinge on my behavior—by making things cost more or less, of course, but
also mostly by making their workings more obvious, giving me a stake, and putting them closer to my fingertips.

When the entire industrial process
is made explicit, when the metrics count for
more than the object they measure,
then GIZMOS become SPIMES.

The Product-Consumer technosociety had a rather simple, linear set of relationships between Consumers and manufacturers. That simpler linearity was composted and subsumed by the GIZMO technosociety. As an End-User today, even a wine bottle will deliberately lure and reward me for becoming a stakeholder.

In a SPIME technosociety, we've advanced into yet another situation, where the core activities involve negotiations over the nature of my stake holding. This activity I call "Wrangling." A SPIME technosociety will be composed of "Wranglers." Effective design helps Wranglers to Wrangle better.

4.

THE PERSONAL IS HISTORICAL

It's time to explain why a novelist would put up with cognitive loading and opportunity costs, just so he could write about design. Allow me to confide in you.

Basically, this is a personal legacy.

My late father (who was also known as "Bruce Sterling,") was an engineer. He was a globetrotting plant manager for an oil multinational. During most of his lifetime, my father was a Consumer in a *PRODUCT* technosociety. But he'd grown up on a Texan ranch where farming and hunting were the everyday business of existence. So he personally experienced a wrenching transition.

His engineering career involved building and managing huge refineries seething with complex, hazardous chemistry. Of course this work brought him more wealth, skill and sophistication than he could have acquired as a rural South Texas farm boy. But that success came with a personal cost. As soon as he could afford it, my father bought land. After that, most every weekend, he left his industrial plant to dote on his Texan ranch.

Eventually, he retired from engineering to devote full-time attention to his cattle herds and orchards.

He never made much money at it. However, there's no question that his rural spread was wondrously well-engineered. Agricultural extension agents flocked in from counties around to marvel at the high-performance rationalism that he'd brought to that ancient pursuit of herds and crops. He was a devoted amateur.

In truth, his ranch was never really about ranching. It was his way to find personal integrity in the dizzy transformations of the 20th century. When he brought his adult skills, his adult ways of thought and action, back to the scenes and situations of his childhood life, that brought him contentment.

By his nature, he was an optimistic, can-do spirit, an engineer and a man of deeds. But I never saw him happier than he was with his trees and cattle.

I myself happen to work in the culture industry. I've also worked on ranches, so I understand what it means to dig postholes, stretch barbed wire, bring in crops and herd obstreperous cattle. Farming has little appeal for me.

Instead, I have the same backward-glancing, chop-licking interest in multinational heavy industry that my father did for his ranching.

I fully share my Dad's intuitive conviction that a serious-minded adult life involves doing something else than what I really do for a living. My dad believed that authenticity had something to do with land, crops and cattle.

In my own mental universe, grown-ups work on massive plant start-ups in remote corners of the world. They install newfangled industrial capacity with state-of-the-art hardware. Surely they wouldn't trifle away their lives writing science fiction novels (although I've been doing that for decades, and nobody tells me to stop).

Of course, I myself don't engineer chemical plants, and I never will. But I do write a great deal about technology. That became my theme as an artist. The human reaction to technological change—nothing interests me more. I want and need to know all about it. I want to plumb its every aspect. I even want to find new words for aspects of it that haven't as yet been described.

So, in the due passage of time, my own life came to resemble my dad's—though in shape rather than in substance. In much the way that my dad engineered his farm, I write literature about technology.

My grandfather, my father's father, was a true rancher. He was a man who lived by the whims of Texan land and weather. I never asked my grandfather what he himself pined for—by the time that I identified this family trait, my grandfather had passed on. I'm pretty sure, though, that my grandfather's own nostalgia was for the vanished, six-gun-toting, nomadic life of the free-range cowboy. That true-blue, saddle-tramp lifestyle lasted a mere quarter-century, but American society has been sentimentalizing it ever since.

I would bet that my daughters, in their own adulthood, will have similar feelings about science fiction books. It's not that books will cease to exist—we still have farms, ranches and petrochemical plants, too—but one can see books shrinking into a smaller media niche under a tidal wave of digital interaction. My daughters are not literary creatures. They are Web surfers and gaming fans. So they'll probably consider ink on paper, sold commercially in bookstores, to be the sign of an ivied, contemplative, solidly classical information economy; in a word, for them and theirs, that will be the sort of thing the old folks used to do.

Like my father, I'm engaged in a struggle to assert some sense of personal integrity within the passage of my own lifetime. I'm a child of the oil diaspora, trying to make sense of the powerful forces that uprooted me from local culture, flung me across the planet and made me an autodidactic gypsy. I'm a science fiction writer—but neither "science" nor "fiction" tells me much of use here.

Design, however, is rather eager to discuss the matter.

By no accident, American design and American science fiction both date to the 1920s. In the visionary work of, say, **Norman Bel Geddes**, with his gargantuan transatlantic airliners and inhabited Hoover dams, it's easy to spot a science-fictional sensibility that hasn't yet been caged and tamed. In their youth, both design and science fiction centered unashamedly on wonder, speed, and spectacle.

Their deepest and more lasting commonality is their fierce love of gadgetry. Design loves the glamorized object; while science fiction loves rayguns, robots, time machines, and rocketships—imaginary objects whose one great unity is that you, the reader, are never going to own one. There is no danger of science fiction's pet gadgets becoming obsolete and disenchanting you. The tide of wonder never ceases for technologies that remain fantasies.

Suppose, however, that you become genuinely interested in gadgets—not as symbols of wonder to be deployed as sci-fi stage props, but as actual, corporeal physical presences. It may dawn on you that you are surrounded by a manufactured environment. You may further come to understand that you are not living in a centrally planned society, where class distinctions and rationing declare who has access to the hardware. Instead, you are living in a gaudy, market-driven society whose material culture is highly unstable and radically contingent. You're surrounded by gadgets. Who can tell you how to think about gadgets, what to say about them—what they mean, how that feels?

Science can't do that. There is no such scientific discipline as "Gadgetology." If you want to write effectively about gadgets, you must come to terms with design. And it pays to make that effort of comprehension, because, in science fiction, as in any kind of fiction, it improves the work remarkably to have a coherent idea of what you're talking about.

> ### Design is not science.

Design has few universal scientific laws to offer us. You can ponder many a design text without ever finding a quadratic equation, a testable hypothesis, or an experimental proof. But design thinking affected my science fiction profoundly, and justly so.

I've been writing "design fiction" for years now. Design fiction reads a great deal like science fiction; in fact, it would never occur to a normal reader to separate the two. The core distinction is that design fiction makes more sense on the page than science fiction does. Science fiction wants to invoke the grandeur and credibility of science for its own hand-waving hocus-pocus, but design fiction can be more practical, more hands-on. It sacrifices some sense of the miraculous, but it moves much closer to the glowing heat of technosocial conflict.

In order to involve one's self in the design world, one has to blunder somehow into the design subculture. This first happened to me in 1989, in Nagoya, Japan. I was invited to an international conference of societies of industrial design. I never understood who wanted me to go there, or what they thought I might contribute to the discourse, but of course I went. I never go to Japan without enjoying myself hugely.

Furthermore, it's one of the perks of the science fiction profession that one never really "belongs" anywhere. People in other walks of life are always surprised to meet a self-declared science fiction writer.

There's a script for it:

"What do you do?"
"I'm a science fiction writer."
"Really?"
"Yes."
"What are you doing *here*?"
"Oh," one likes to drawl, "this way-out scene of yours is very science fictional!
Yeah, really! I'm, uh, trend spotting here!"

The first industrial designer I ever met, one **Tucker Viemeister**, did not bother to ask me these things. Instead, over a couple of whiskies at the top of a Japanese skyscraper, **Mr. Viemeister** simply began, in his uniquely orthogonal fashion, to get me up to speed with the nature of his milieu. When I tell other designers that the first designer I met was **Tucker Viemeister**, they can only nod knowingly. The truth is that this **Viemeister** character was and is a designer's designer, one of those luminous, Eames-like beings who do not so much pursue design as embody it. I didn't fully grasp that in 1989, but I immediately recognized that I was in the presence of someone culturally radioactive.

For an SF writer, any guy like this one was top-notch material. I needed to know all about him and all his friends. So I learned.

"The task of the artist is to create, not to talk," as Goethe supposedly said (among his many fits of talking). I'm a creative artist myself, but in the design world, I am liberated in a way that I can never be in fiction. In design, I am not a creator. There, I'm just a talker. I have no design talent. I feel no burning desire to design anything much. Still, my interest in design is personal, sincere and long-term. So, by my nature, I'm suited to be a design critic.

So I became one, and wrote about it more and more often. I've become a design fan of the purest dye. However, unlike most design fans, I'm not a fan of designers. I quite like designers—I never met a designer I didn't like—but I'm not in the design game.

By my nature, I'm a fan of design *teachers*. I'm an earnest devotee of design gurus. Design gurus like to grumble that they are rarely taken seriously as public intellectuals, but since I'm a science fiction writer, these shibboleths bother me not one whit. You bet they're important and persuasive public intellectuals; because I read their work, and I'm swiftly motivated to throw away startling sums on machinery, shelter, clothing, and working tools. If that isn't a potent influence, what is?

People sometimes imagine that designers are thorny, arcane, unapproachable divas. Yes, sure, but that's all for show. Once one gets to know designers, one quickly finds that designers have remarkably low-key egos, especially compared to us authors. By their nature, designers are accommodating problem-solvers. Their basic instincts lead them to unsnarl social embarrassments with graceful efficiency. So designers make ideal hosts. Their events always have great signage and pretty brochures. They are clever, quick studies and they understand arcane jokes. I dote on them.

The second designer I met was a gentleman named **Mike McCoy**. I admired the handsome office chair **Mr. McCoy** was sitting in, and he informed me that he himself had designed it. He then rose from the chair, turned it upside down on the spot, and identified and described all its moving parts. I was awed. (Really, anyone of sense should be awed by such accomplishments.) I went home and bought that chair, a major reason why, despite my years of ceaseless typing, I'm never troubled by lumbar ailments.

Design and science fiction are sister occupations, but it's vital to make one important distinction. Compared to designers, science fiction writers are visionary cranks. Crankhood isn't so much an occupational hazard as a core requirement of my trade. Science fiction just doesn't work as a literary genre without a strong whiff of the visionary. Effective science fiction always has some kind of burning, subterranean agenda, on the verge of bursting out of control.

With Jules Verne, it's a frustrated sailor stuck in a bourgeois attic. With H. G. Wells, it's a leftist revolutionary writing thrillers to make his rent; with H. P Lovecraft, it's mind-blowing cosmic vistas from a put-upon shabby gentility. Science fiction is not about the freedom of imagination. It's about a free imagination pinched and howling in a vise that other people call real life.

Writing is burdensome. A mind at full ease with itself would not need to slither onto a page; a serene mind would not need to speak its mind. When I read other writers, I know enough about writing that I am always aware of the compositional struggle; so whenever I read good writing, I perceive hard work. I feel pangs of collegial sympathy. But when I see good design, I don't suffer along with the laboring designer; I know he's working hard, but I just feel happy. I am free to feel the innate joy of human creativity, and really, I'm grateful for that. Design gives me joy. That's not a small matter in life.

However, one can't merely trifle in the lives of others; if you're a design dilettante, you do run the hazard that, somehow or other, you may actually design something. This mishap once happened to me. Some eccentric European artists once commissioned me to design a lamp.

Of course I accepted that commission; how could I refuse? The mere fact that I lack design talent was only, well, an interesting design constraint.

I have no innate feeling for form. I can't draw, I can't visualize well in three dimensions, I have no keen awareness of detail, I have poor experience with material affordances, and when it comes to handiwork, I could easily cut my thumb off slicing a bagel. Frankly, I'm just no good.

However, I don't have much trouble thinking like designers think. In particular, I can extrapolate; if you show me A, B, and F, I can leap to Z in a trice. So I took a leaf from the practice of two of my favorite contemporary designers, **Laurene** and **Constantin Boym**.

These **Boym** worthies have the remarkably science-fictional habit of going out to hardware stores and repurposing common objects for bizarre uses never intended by their manufacturers. If you like *Shaping Things* ISBN 0-262-69326-7, you would probably like their book, *Curious Boym* ISBN 1-568-98353-0, much better. Not only is it wittier and cleverer than my book, but the **Boyms** also seize the opportunity to settle scores with any number of deadbeats who rooked them in their design practice, which is hysterically funny.

So I followed in the **Boyms's** pioneering footsteps and I created a lamp out of common, everyday cable clamps. Cable clamps are nonconductive plastic knickknacks commonly used to control thick, unruly clumps of power cables. These clamps are cheap and durable, and they would serve well hold the power cables of a whole office-full of lamps. So (I reasoned) why not cut to the chase and make the lamp itself out of the clamps?

As it happens, these toothed and jawed cable clamps have some interesting sculptural qualities. So it's fairly simple matter to superglue them into a weird, rippling, toothy, anemone-like construction. I didn't make the best possible lamp that could be made out of cable clamps. But since nobody else would think to try it at all, I created a unique object that looks like no other lamp on earth. My lamp looks extremely science fictional without being in any way derivative of previous efforts. So, well, the project worked. At least, the clients were pleased: mission accomplished.

I fondly imagined that this lamp of mine was going to be durable, practical, democratic and cheap. Of course, **Harry Bertoia** also thought that about his Diamond chair made of simple steel wire, and **Marcel Breuer** thought that about his Wassily chair made out of simple steel tubing, and the **Eames** thought it about their Potato Chip chair made of simple glued plywood. My lamp made of plastic clamps is not cheap at all. It is a hand-glued *recherché objet* that costs an arm and a leg in toney French art galleries.

Of course I was intellectually aware of this conundrum in design, but experiencing it myself has made me a better critic.

I showed my lamp to **Tucker Viemeister**. He looked at it silently for about thirty seconds and said that it was

"good."

I came full circle, right there.

5.
METAHISTORY

How do people know what to expect from their things? Every culture has a metahistory. This is not the same as their actual history, an account of places and events. A metahistory is a cultural thesis on the subject of time itself. Metahistory is about what's gone by, what comes next, and what all that is supposed to mean to sensible people.

As a science fiction writer I find these social constructions of particular interest. How do people come to grips with the future? How do they think about futurity? How are those judgments made and how do we alter those judgments?

A culture's metahistory helps it determine whether new things are appropriate, whether they fit into the trajectory that is considered the right track. For instance, if you happen to be an Egyptian pharaoh, it makes perfect sense to assemble the populace in the off-season to create huge granite and limestone time-machines for your posthumous existence. That is by no means an exotic whim; there is nothing romantic or visionary about this critical social project; building pyramids for the Pharaoh's future demise is common-sensical and entirely decent.

We moderns behave in much the same buoyant, unthinking way when we disinter fantastic volumes of coal and crude oil, set fire to them, and export the smoke into the sky. This was critical to our sense of progress once; we've yet to understand that it is radically harming our ability to go on.

To understand metahistory is vital,
but not sufficient.

Now let's step up one analytical level. What is the future for history itself? What metahistorical thesis comes next? Can we speculate effectively about future metahistories that will be created and believed by future cultures?

"Future metahistories" would be grand narratives about time that are as yet inconceivable, and can't as yet be held by anybody. We may not be able to predict or describe future metahistories, but to judge by the long-established historical trends, it's entirely clear that there will be some. Metahistories do exist; they come to exist; they fade from the cultural landscape; they have limited life spans.

And yet, there has never been a metahistory that can recognize itself as provisional. Grand ideas about time always consider themselves to be somehow time proof. All around us we see obsolescence—but our ideas about obsolescence are not supposed to obsolesce.

Can we transcend this failure of insight? Can we make room and offer a cheerful welcome within our own metahistory, for unborn metahistories whose time is not yet here? Can we allow ourselves to understand that our deepest ideas about existence are themselves mortal formulations?

Why do we want to make this effort? It's because a metahistory is the ultimate determinant of the shape of things. It's through metahistory that people come to realize that new things are proper things. New objects that can fit into a metahistorical context are seen as progressive advancements. Otherwise they are considered alien impositions or odd curiosities.

It may be that any attempt to tamper with metahistory is inherently wrongheaded. Karl Popper (the advocate of paradigmatic thinking in the philosophy of science) held that "oracular philosophy" is a "poisonous intellectual disease." According to Popper, some people drink too deep at the predictive well of metahistory, imagining that they can create utopian *gesamtkunstwerks* and engineer some new kind of humanity. These dangerous meddlers will find themselves fatally led toward "interventionism." Popper considered interventionism to be inherently evil, for interventionism is the mortal enemy of a looser, more open model of society that can allow open dissent and openly redress its errors.

It follows that oracular philosophers, in their well-meaning attempts to engineer society, will find themselves corrupted and turned sinister, forced to strangle human freedom and deny the unforeseen.

I'm inclined to agree that imposing oracular philosophy with guns and gulags is diseased. The USSR was the number-one example of a totalitarian, interventionist society. It failed, so it's certainly not one any more. The Russians gave up their doctrine that Marxism-Leninism made them the "avant-garde of mankind."

So the Russians are now free of interventionist disease—but Russian society today is vanishing. Death rates have soared, birth rates have plunged, real diseases are laying waste to the populace, and Russia is in a catastrophic demographic decline.

There seems to be no easy explanation as to why this would happen to people in an industrialized state, at peace, with huge natural resources. One would think that their hearts were broken; that they had no place to turn to face the coming day and had lost hope of finding any; and that this metahistorical collapse is proving fatal to them and their culture.

> The premier argument
> for metahistorical intervention is that
> the status quo will kill us.

Let's consider the main challenges that threaten our civilization's continued existence. That would be climate change first and foremost. There is also deforestation, the impending death of jungles and coral reefs, over-fishing, soil erosion and salinization, emergent and resistant plagues, depletion of the world's fresh water supplies, exhaustion of fossil fuels, a bio-accumulation of toxins in water, food and soil, in human bloodstreams and human flesh, plus increases of crashes of the world's population, and the possible use of weapons of mass destruction.

With the exception of WMD, which is simply genocide in a can, these are all slow crises cheerfully generated by people rationally pursuing their short-term interests, from within a metahistorical framework they have yet to mentally transcend. Today's open and democratic societies are also unsustainable, so they are about as good at generating these crises as any other societies.

We shouldn't gloomily imagine that all possible threats to society are somehow all our own doing, either. The planet is inherently hazardous. There are rare but real threats such as asteroid strikes, super volcanoes, vast tsunamis, the reversal of the Earth's magnetic field... A society determined to thrive in the long term would need to keep a wary eye on these matters. It should be prepared to deal with such challenges on the scale and with the energy that a threat of that scope would require. A society that abjures intervention on principle can't do that.

> A society that can't sustain itself may have strong ideas about its metahistory, but objectively speaking it has no future.

What is needed is the energy for effective intervention without the grim mania of totalitarianism. We need to take action without any suffocating pretense of eternal certainty. So we need a new concept of futurity whose image is not the static, dated tintype of the past's future. We need a dynamic, interactive medium—we need to invent a general-purpose cultural interface for time.

Metahistories to date have had the static character of a sacred oracular text. What we need to invent is something rather more like a search engine. We need a designed metahistory.

History is never a deterministic certainty—understood effectively, history is a *basic resource*. We would think of time and futurity very differently if we came to understand that the passage of time could make one rich. It can. Because history is information—information about the people and objects transiting time. The word "information" should suggest not some frozen ideology or timeless gospel, but economic activity. That would be history as business, history as governance, history as symbolic analysis—history etched into the very texture of the technosocial.

Combine the computational power of an INFORMATION SOCIETY with the stark interventionist need for a SUSTAINABLE SOCIETY. The one is happening anyway; the other one has to happen. When opportunity meets necessity, invention takes place. It is a new, 21st century society with a new, progressive form of metahistory.

This is why "SPIMES"—or something partaking of their characteristics – are our era's hopeful children. SPIMES are information melded with sustainability. Without sustainability, information is top-heavy, energy-hungry and heading for a crash; while sustainability is impractical without precise, comprehensive information about flows of energy and materials. A SPIME is a class of objects with the capacity to attend to both.

So a SPIME, understood properly, is not merely the jazzed-up descendant of today's barcodes and ID chips. SPIMES are the intersection of two vectors of technosocial development. They have the capacity to change the human relationship to time and material processes, by making those processes blatant and archiveable. Every SPIME is a little metahistory generator.

A technosociety skilled with SPIMES can maintain itself indefinitely through a machine-mediated exploitation of the patterns of movement of people and things through time.

History is this technoculture's primary source of wealth. As it transits through time, due to the principles of its organization, it will increase in knowledge, capacity, wealth, and power. It has the means, motive and opportunity to sustain itself in the most profound sense of the term.

I now want to make a case about this visionary scheme: it's visionary only in the abstract way that I have been phrasing it. As experienced, it would seem quite bland

and practical. It is happening already on many fronts. It has implications that are governmental, educational, military, industrial, financial, ecological—it is societal, civilizational. And personal as well.

For a society of this sort—we might call it a SYNCHRONIC SOCIETY—history is not "a nightmare from which we are struggling to awake." History is the means by which we wake up. We wake up, and we go about our daily affairs, free of shadows of imminent apocalypse and secure in the objective knowledge that our activities as civilized beings are expanding our future options and improving our current situation. This is how we would interact with time if we human beings were really on top of our game.

6.

A SYNCHRONIC SOCIETY

A SYNCHRONIC SOCIETY synchronizes multiple
histories. In a SYNCHRONIC SOCIETY,
every object worthy of human or machine
consideration generates a small history. These histories
are not dusty archives locked away
on ink and paper. They are informational resources,
manipulable in real time.

A SYNCHRONIC SOCIETY generates trillions of cata-
logable, searchable, trackable trajectories: patterns
of design, manufacturing, distribution and recycling
that are maintained in fine-grained detail. These are
the microhistories of people with objects: they are the
records of made things in their transition from raw material, through usability, to evanescence, and back again to
raw material. These informational microhistories are
subject to well-nigh endless exploitation.

Exploiting this potential successfully is a major opportunity and challenge for tomorrow's design. It is something
never done before, a place where the shapers of tomorrow's

things can develop possibilities unavailable to any previous generation. I call it a metahistorical issue, because that's the best way to summarize it—but when it comes to actually instantiating this trend in real things, real material goods and real immaterial relationships, it will always be a design issue.

Historians won't do it. Designers will. In particular, 21st century designers will do it, because it was not just impossible, but unthinkable, to earlier designers. It is a realm of design opportunity untouched by all predecessors.

This vast digital bulk of trillions of histories is burdensome and even hazardous in some ways. It requires huge resources in bandwidth, processing speed and storage. There is every reason to think (based on firm 50 year trends) that those resources will exist. Since they will also allow new forms of behavior and new relationships between human beings, the environment, and their objects, they are intensely valuable.

Sustainability is never a static goal. It can only be a process. Previous ideas about "sustainability" are not and will never be tenable. *A small, beautiful, modest, handcrafted society, living in harmony with its eco-region, relentlessly parsimonious in its use of energy and resources, can't learn enough about itself to survive.* In its bucolic quietude, it may appear timeless, but the clock is ticking for it as it does for all societies. It can avoid many conventional threats by abjuring large-scale, clumsy technologies, but modesty doesn't make one invisible. That society isn't *keeping track*—in its loathing for industrialism, it forfeits far too much command-and-control over its physical circumstances. *Its bliss is ignorance.*

A truly sustainable society has to be sustainable enough to prevail against the unforeseen. The unforeseen, by definition, can't be outplanned. This implies that serendipity is necessary. We can't know what we need to know; so there need to be large stores of unplanned knowledge.

There is the known, the unknown known, and the unknown unknown. When the unknown unknown comes lurching to town, you have to learn about that comprehensively and at great speed. Generating new knowledge is very good, but in a world with superb archives, accessing knowledge that you didn't know you possessed is both faster and more reliable than discovering it.

This is the new form of knowledge at which a SPIME world excels. It is not doctrine, but the school of experience—not reasoning out a solution a priori, but making a great many small mistakes fast, and then *keeping a record of all of them.* This is where the 21st century has a profound oracular advantage over the intellectual experience of all previous centuries—it can simply *search the living daylights* out of vast datamines of experience, at the press of a button.

The ability to make many small mistakes in a hurry is a vital accomplishment for any society that intends to be sustainable. It's not necessary that every experience be sensible, logical or even sane—but it's vitally important to register, catalog and data-mine the errors.

In the world of design, the term for this is "rapid proto-typing." Rapid prototyping is a form of brainstorming with materials. It's not simply a faster way to plunge through older methods of production, but a novel way to manage design and production. By previous standards, it looks as if it is profligate, that it "throws a lot away"—but with bet-ter data retention, "mistakes" become a source of wealth. Rapid prototyping seen in depth is an "exhaustion of the phase space of the problem"—it isn't reasonable, thrifty or rational, but it has the brutal potency of a chess-play-ing computer.

Designers brainstorm. It's not reasonable to brain-storm. A brainstorm works anyway, because the point of brainstorming is escaping "reasonable" constraints. A brainstorming session fails if remains too reasonable. Brainstorms are about generating fresh, effective ideas from outside some particular paradigm.

As designer **Henry Dreyfuss** used to say, a brainstorming session will produce three good ideas at the cost of 97 bad ones—a cost, said **Dreyfuss**, that had to be borne as the price of the three good ideas. What is intellectually dif-ferent about the 21st century is its improved mechanical ability to winnow out the three good ones among the 97 bad ones—and to keep the 97 bad ones around so that we needn't do them again.

A society with SPIMES has design capacities closed to societies without them. Since they are so well documented, every SPIME is a lab experiment of sorts. In older days, if an object was radically re-purposed by some eccentric, this data would be ignored or lost. A SYNCHRONIC SOCIETY is in a splendid position, though, to adopt and refine these innovations. A mass produced object can be compared to a grazing cow, while the same basic object, when SPIMED, becomes a scattered horde of ants. Each ant pursues a different trajectory and therefore covers a broader spectrum of technosocial possibility.

A world with SPIMES, in other words, can make and correct missteps faster than earlier societies, and with less permanent damage. SPIMES are a digital mob of tiny, low-cost advantages and mistakes. A SYNCHRONIC SOCIETY can study history in more depth—farther into the past, farther into the future—but also operates in more breadth. Instead of researching new solutions from a standing start, it has a new capacity to digitally search out solutions within the existing data field: every SPIMED object has generated a little puddle of experience.

A SYNCHRONIC SOCIETY has a temporalistic sensibility rather than a materialistic one. It's not that material goods are unimportant—materials are critical—but material objects themselves are known to be temporary, obsolescing at a slower or faster pace. A SYNCHRONIC SOCIETY conceives of its objects, not as objects qua objects, but as instantiations, as search-hits in a universe of possible objects. Embedded in a monitored space and time and wrapped in a haze of process, no object stands alone; it is not a static thing, but a shaping-thing. Thanks to improved capacities of instrumentation, things are no longer perceived as static—they move along a clocked trajectory from nonexistence to post-existence.

How do we learn to think in a SYNCHRONIC way? Through using **MACHINES**. Genuinely radical changes in the human conception of time are not caused by philosophy, but by instrumentation. The most radical changes in our temporal outlook come from technological devices, tools of temporal perception: clocks, telescopes, radiocarbon daters, spectrometers. It was through these instruments that we learned that the universe is 13.7 billion years old, that the planet is 4.45 billion years old, that our species is some 200,000 years old. Compared to these mechanically assisted vistas, all previous human notions of time are parochial.

Then there are sensors, which do not merely measure qualities, but measure changes. Sensors that can measure and record. Sensors for changes in temperature. Sensors for changes in moisture. Sensors for changes in light. Sensors for changes in magnetic fields. Sensors for changes in chemical exposure. Sensors for the changes wrought by microbes and pathogens. Sensors for changes in chemical exposure. And clocks, cheap, accurate, everywhere, measuring changes in time.

FIGURE 2

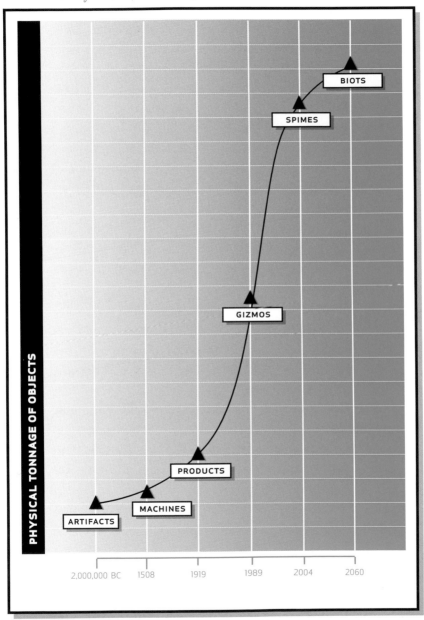

The Human Engagement with Objects
Artifacts, **MACHINES**, *PRODUCTS*, SPIMES, **Biots**

A SYNCHRONIC SOCIETY is fascinated with ideas about progress and advancement. But it doesn't want society to move in lockstep unison into some prescribed direction; it wants to generate the potential to move in effective response to temporal developments. A civilization cannot outguess all eventualities, so it has to cultivate capacity, agility, experience, and memory.

A SYNCHRONIC SOCIETY would view human beings as process: a process of self-actualization, based not on what you are, but what you are becoming.

The value judgments of a SYNCHRONIC SOCIETY are temporalistic. "Do we gain more time by doing this, or less time?" Does this so-called "advancement" increase, or decrease, the capacity for future acts?

Consuming irreplaceable resources, no matter how sophisticated the method, cannot mean "progress," judged by a SYNCHRONIC perspective. Because to do so is erasing many future possibilities; it is restricting the range of future experience.

Constructing hydrogen bombs was once a highly sophisticated technical effort. Huge bombs might even be politically or technically necessary in the midst of some gigantic, all-or-nothing crisis (say, huge bombs for use against an asteroid in imminent danger of smashing the Earth). From a SYNCHRONIC viewpoint, though, creating and storing world-smashing super weapons can't possibly be judged an "advancement." It's a blatant, future-wrecking hazard, no matter how clever it is, or how

difficult or costly to do. The use of hydrogen bombs fore-
closes practically every other act of future development.

A SYNCHRONIC SOCIETY sets high value on the
human engagement with TIME. We human beings are
time-bound entities. So are all our creations. We cannot
think, analyze, measure, prove, disprove, hypothesize,
argue—love, suffer, exult, despair, or experience a word-
less rapture of mystical faith—without a flow of TIME
through our flesh. So we are not objects, but processes.
Our names are not nouns, but verbs. Our existence does
not precede time or postdate time—we personify TIME.

If we accept this philosophizing, certain implications
follow. When someone's lifespan is curtailed, this fore-
closes that person's future experience. So, living a long
time in full awareness of one's circumstances is a praise-
worthy act. Blowing yourself up and killing those around
you in pursuit of a supposed eternal reward must be close
to the apex of wickedness.

Temporalistic thinking is a moral worldview. A society
with declining life expectancy is clearly retrogressive. A
society with a high infant mortality rate is maladjusted. A
society riddled by plagues, diseases, resistant and emergent microbes, and environmental illnesses is decadent.
Societies facing these blatant danger signals need to
frankly come to terms with their decline. People of good
will in such a society should frankly recognize and publi-
cize its failings, and take appropriate remedial steps.

Or so one imagines a SYNCHRONIC SOCIETY moralizing.

Of course, this is speculative. Even if we did effectively think and act in such a way, it's unlikely that we would ever use such a cumbersome label as SYNCHRONIC for our sensibility. But we could act and think that way if we wanted to do so; there's nothing much stopping us from doing it right now.

I suspect that we are quite close to thinking this way, and what I am describing here is a clumsy, old-fashioned prognostication for a way of life and thought that will someday be so common as to be banal. A sensibility like this sounds rather exotic in the TIME in which I write this. It would make a great deal more sense, however, in a future society with a burning awareness of environmental crisis, where the majority of the population is well-seasoned, elderly, adept with media and surrounded by advanced computation. That is a very plausible description of the mid-21st century cultural scene. They would read a book like this and laugh indulgently—but they would read many other books of our period, and wonder in shock what on earth those people had been thinking.

We're in trouble as a culture, because we lack firm ideas of where we are in time and what we might do to ensure ourselves a future. We're also in trouble for technical and practical reasons: because we design, build and use dysfunctional hardware.

7.

THE RUBBISH MAKERS

Hardware has no value judgments. Hardware has no faith and convictions. Hardware is not a moral actor. Hardware is our method for engaging with the grain of the material. We human beings have never done that with genuine efficiency and elegance. We're still learning. Now there are over six billion of us, and the consequences of past misbehavior with hardware are all around us.

To understand hardware, we need to understand hardware's engagement with T I M E .

Hardware is prehistoric. Hardware is prehuman. Technology is older than people. In the long and intimate relationship between humans and objects, objects are the senior partner. Tools are probably older than speech.

We humans are what tools made of us. The human body, human perception, human intelligence, they are all the outcomes of two million years of hominids interacting with hardware.

When human beings first appear on the landscape of time, some 200,000 years ago, humans appear breaking rocks and using fire. There were, and are, no humans so primitive that they lack these technical accomplishments. That's because they're not human accomplishments.

Those are *prehuman* accomplishments that we humans inherited from a previous species.

Man is "man the toolmaker," which is to say that human beings excel at a deeply attentive mental and physical engagement with artifacts. No other species begins to rival us in this intensely human mode of being. Some animals do occasionally use tools, in immediate, spontaneous ways, but animals lack any sustained interest in creatively tinkering over extended periods of time.

Animals can't design. Apes will fling objects, but humans will throw objects, practice throwing them, and refine the grain of the material so that the thrown object throws better. Humans have evolved an innate capacity to shape things: they have habits, customs, bodies of transferred know-how. Humans create infrastructure. Humans get far better at interacting with objects than any animal can ever manage; and since humans are also capable of abstract analysis, they are also better at getting better. Humans have technosociety.

What we know about prehistoric humans comes mostly from their things. Prehistoric peoples left us no documentation, since they were pre-literate. However, they left many things that they shaped, then discarded or lost. Occasionally, abandoned and forgotten Paleolithic artwork is found, deep in caves or in lonely deserts. Sometimes, we discover fragments of their bodies.

If we were to judge ourselves by the efforts of ours that survive the passage of time, we'd be best described as Man the Rubbish Maker. We've been polluting since before we were human.

Chipping rocks into tools is a messy, haphazard process. When archeologists investigate ancient rock foundries, they always find vastly more rock waste than they ever find tools. Rock waste is the earliest form of pollution. It is an unsought, useless, and hazardous externality to a technological process.

Paleontologists have found flint-knapping workshops two and half million years old where the stone junk is still sharp enough to slash an unwary hand. The rock waste from prehuman workshops has lasted for geological time-scales. Their speech, their culture, their beliefs about time and futurity—that's gone like an exhaled breath.

Any attempt to shape things, any physical act of material engagement, involves a transfer of energy, a friction, a transfer of atoms, an emission of photons... there is always some subtle re-shaping. Some is intentional and useful. Much is not.

> Entropy requires no maintenance.

Because we humans enjoy things and use things, our favorite things wear out quickly. Pollution is not subject to our consumption. So pollution tends to persist, while the useful tends to wear out. Nature can subsume many forms of human pollution, but even nature misplaces some valuable resources. That's why we have oil and coal. Oil and coal are natural pollution. Oil and coal are sources of biotic energy that the biosphere did not efficiently recycle. Fossil fuels are necrotic energy.

It is difficult to deal with rubbish. Humans have always failed to deal with our trash as we made it. The role of trash is therefore exalted over the longer term. Civilizations collapse, but their ruins are a byword. Trash is always our premier cultural export to the future.

Our ancestors often spoke about posterity and immortal fame. What they would most like us to remember about them is generally gone. Since our values differ from theirs, we often can't even praise them for what they considered their virtues. What we do receive in plenty from past people is middens and refuse. Middens and refuse are objects so deprived of value that no human being tries to shape them. We have plenty of rubbish. It lasts for thousands of years.

We ourselves, the people of our own time, are leaving ultra-long-lived rubbish of two distinctly innovative kinds: atomic-age radioactive waste, and space age junk satellites wandering in distant orbits. Radioactive waste has left an unmistakable mark of atomic tinkering everywhere on this planet, from the polar snows to the sediments of the abyss. Future archeologists who captured an abandoned spacecraft could learn an amazing amount about the Space Age. Spacecraft are sophisticated, multiplex artifacts crammed full of involuntary

FIGURE 3

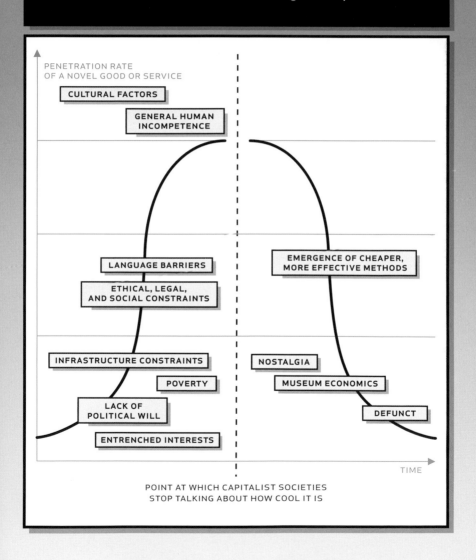

clues about our strenuous efforts at command and control. Our use of spacecraft and nuclear fission—no, not the glorious technical accomplishments, but the ageless pollution that they generated—will be evident for geological time-spans.

Let's consider how objects commonly behave in the passage of time.

The double S-curve is the standard trajectory of objects passing through time. Objects arise from means, motivation, skill, and material opportunity. Then they diffuse through a population. Objects do not *evolve* or improve automatically; they enter subcultural niches where they undergo various and sundry developmental variants. "Form follows function," but objects function in a technosocial context.

Businessmen have renamed the doubled S-curve from their own point of view. The fast money and big profit margins are in the first fine arc upwards: the *Rising Star*. Steady money and blue-chip business power are in *Cash Cow*. There is no revenue to be found in the earliest part, called *Question Mark* (Research and Development) or in the final phase, *Dead Dog* (Obsolescence).

How does all this happen? It isn't magic. People do it. Some people are much better at it than others.

8.

THE STARK NECESSITY OF GLAMOR

Let's consider, in some detail, an historical actor who is a galvanizing figure in a process of technosocial development. An excellent candidate would be **Raymond Loewy** (1893–1986), the self-proclaimed "Father of Industrial Design."

We owe to **Raymond Loewy** the particularly useful acronym **MAYA**, or, **Most Advanced, Yet Acceptable**. This formulation is the key to the Loewy oeuvre. **MAYA**, according to **Loewy**, is what industrial designers are supposed to do with their skills, for their clients, and to the world. Designers create objects, products, processes, symbols that anticipate the future. However, these innovations can also be metabolized on a broad scale by society in general.

It just will not do to settle for the one activity or the other. **Most Advanced** would be ivory-tower scientific researchers. **Yet Acceptable** would be crass mass manufacturers. A designer is neither **MA** or **YA**, but **MAYA**, with all that implies. He is not compromising; no, he is synthesizing! This is not a lack of integrity on a designer's part, but the very source of integrity.

Designers mine raw bits of tomorrow. They shape them for the present day. <u>Designers act as gatekeepers between status quo objects and objects from the time to come</u>.

Designers are not merely making things look attractive to purchasers. One can do that, and it's a valuable skill, but that's not true to **Loewy's** spirit of design—because there's no Advancement.

For instance, consider a man in Santa Fe, New Mexico, who makes a handsome adobe air-conditioner housing in the local "Santa Fe Style." By the **Loewy** standard, this stylist is not a true industrial designer. He is merely tactfully disguising the air conditioner (a technological innovation) while indulging his clients with some retro decor (an adobe shell). **He'll likely stay in business, because he is effectively serving the weird need of Santa Fe locals to preserve the coherent visual stylings of their museum economy.** But nothing has in fact happened in an industrial design sense—he's merely applied a muddy cultural patch over a technological incongruity.

If another Santa Fe designer thinks creatively and crafts an air conditioning system out of New Mexican adobe—(like, say, a wettable, porous ceramic heat-exchanger with perhaps a nifty subterranean heat pump)— then she would become a laudable wizard of **MAYA**, a genuine designer fit to take the technosocial stage with **Raymond Loewy**.

In order to thrive, an industrial designer has to comprehend the ins and outs of the intricate, treacherous bargaining processes involved in an advancing industry. Since we have our own newfangled ideas about what **ADVANCED** might mean (and they differ profoundly from **Loewy's** period notions), let's first unpack the rest of his **MAYA** aphorism: those apparently simpler terms, **MOST**, **YET**, and **ACCEPTABLE**.

The term **MOST** implies that there is not merely advancement per se, but continuous *grades* in advancement. These jolts up the S-curve can be finely judged, oiled and adjusted by a fully briefed, able design professional.

Then there is the interestingly temporalistic term **YET**. **YET** implies that certain people who lack professional judgment will *resist* the advancement. There is an innate cultural friction here. The world does not always beat a path to the better mousetrap. Intelligent designers can create products embodying (as **Loewy** puts it) "logical solutions to requirements" which "express beauty through function and simplification." However, these triumphs of **Loewy's** design craft cannot succeed in public without some brisk combat in a culture war.

This is a given. It is part of the profession. If there is no counter-reaction, it means that there has been no action; no change of substance can have occurred. Many shadowy and vaguely sinister forces have already "conditioned" the public into accepting customary, badly designed items as "the norm." These reactionary forces stand in the way of a designer's logic achieving that moment of **YET**. These forces must be overcome.

YET is not a passive waiting for the clock. **YET** is a vigorous battle for mind-share.

"Success finally came," wrote **Loewy**, "when we were able to convince some creative men that good appearance was a salable commodity, that it often cut costs, enhanced a product's prestige, raised corporate profits, benefited the customer and increased employment."

"Success" came from convincing capitalists—"some creative men"—to give **Loewy** access to the means of production, so that he, **Raymond Loewy**, has the power to alter the shape of things by bestowing prestige, customer benefits, and a good appearance upon them.

This success is not to be found in **Loewy's** brisk reshaping of some backward objects. **Loewy** has many rational, logical arguments to offer his "creative men" in business, but it isn't the reason or the logic that constitute **Loewy's** success. **Loewy** knows full well that he can manage the reshaping, for **Raymond Loewy** is that rarest of things, a trained engineer with an exquisite sense of taste. The success, the moment of **YET**, finally comes from—the success *is*—**Loewy's** inculcation of conviction. He didn't merely reason and logic-chop at clients; he convinced them.

Once that happens, **MAYA** moves with a heave into the next notch up the S-curved ratchet. **Loewy** can then judge that notch and design well for it.

The customers—(those numbed pawns of reactionary false consciousness)—are even more impervious to logic and reason than those "few creative men in business." The customers cannot be harangued with facts in a board-room. The customers must be seduced.

This led **Loewy**, one of the design profession's genuine pioneers, to espouse a grandiose, coruscating lifestyle reeking of transatlantic European chic (even though he was, yes, an engineer). **Loewy**, the Father of Industrial Design, made himself, in a word, **designery**.

Being **designery** is not an affectation. Being **designery** is how one manipulates **MAYA** in public. Being **designery** is what one does, as a practical measure, in order to overcome the reactionary clinging to the installed base of malformed objects that maul and affront the customer. What cannot be overcome with reason can be subverted with glamor. That's what design glamor is for.

Raymond Loewy generated a great deal of professional ballyhoo for himself and the design profession generally. For instance, **Loewy** didn't mind assuming the public credit for the work of his collaborators and subordinates. In the arts and sciences, such behavior is reprehensible, but in a battlefield of **MAYA**, a reputation for omnicompetent genius is a valuable crowbar to jam into the door of the **ACCEPTABLE**. If **Loewy** is publicly seen as a supreme visionary, then his collaborators and subordinates will get more work. This means a greater field in which to exercise their reshaping efforts. Expanding the **MAYA** pie takes priority over the just distribution of the pie. That may be deceitful, but a battlefield is full of feints and stratagems.

It should not to be imagined that **Raymond Loewy** saw a case of narcissistic personality disorder when he shaved himself in the mirror. **Loewy** understood a promotional image, for he was an engineer who also been a Macy's window dresser and a fashion illustrator.

Loewy's period colleagues and rivals, **Henry Dreyfuss** and **Norman Bel Geddes**, were veterans of Broadway stage design. For **Bel Geddes** and **Dreyfuss**, being **designery** was a brilliant mutation of being "theatrical." Actors have ego, but great actors have great craft. Every ham may love the limelight, but great thespians will let the role flow right through them. They don't merely pretend. They embody.

Designeriness is not the core of design. Glamor is an epiphenomenon. But whenever designers cease to be flamboyant, the boom of their artillery ceases. A conspicuous lack of charlatanry and pretension means that little is happening in the designer's cultural battlefield.

Any designer subjected to fame and commercial success will find himself or herself becoming **designery**, almost by reflex. Designers are ingenious and adaptable by temperament. They quickly find that flaunting designery attitudes is the quickest, most efficient form of designer public relations. Being **designery** cuts elaborate discussion short. It gets useful results. It scares up business. It increases capacity, buries the fallen and prepares the next campaign. That's why glamor is a stark necessity.

Henry Dreyfuss was nowhere near so waspish and fussy as **Raymond Loewy**, but **Dreyfuss** sported a custom-made brown business suit in a corporate world of pinstriped blacks and navy blues. **Henry Dreyfuss** was "The Man in the Brown Suit," i.e., **Henry Dreyfuss** thoughtfully embodying the public role of "**Henry Dreyfuss**." **Dreyfuss** wore brown even when designing no-nonsense tractors, thermostats, and the war situation room for the Joint Chiefs of Staff. **Dreyfuss** even had a custom-tailored brown evening suit for formal wear at the theater.

ADVANCEMENT and **ACCEPTABILITY** have to be created by capturing the public imagination. One cannot buy a kilo of **ADVANCEMENT** or rent a liter of **ACCEPTABILITY**. These immaterial barriers have to be budged through eye-catching acts of inculcated conviction.

Progressivism versus conservatism is culture war. People who win or lose a culture war don't merely act as if they won or lost a culture war; they genuinely win it or lose it. It's always war—and if it's not magnificent, then designers are losing.

ACCEPTABLE is a time-bound condition. Standards of **ACCEPTABILITY** transform over time. They are never absolute. Any absolutely **ACCEPTABLE** object would have no grain of social resistance. There would be nothing to quarrel over, nothing to discuss, no vector of improvement. An absolutely **ACCEPTABLE** object would be invisible. Having fully "solved" its design "problem," and achieved a perfectly classic grace and functionality, it would offer no drawbacks and no scope for human invention. A fully **ACCEPTABLE** object would be an extremely mysterious thing, ghostly and uncanny, beneath conscious notice, almost beyond thought.

No material thing can ever achieve full and utter **ACCEPTABILITY**. People are too ductile to have their problems solved. People are not parameters for design problems. People are time bound entities transiting from cradle to grave. Any "solved problem" that involves human beings solves a problem whose parameters must change with time. A "thing" is no more stable than the humans who cherish it. Properly understood, a thing is not merely a material object, but a frozen technosocial relationship. Things have to exist in relationship with an organism: the human being.

Things can lose their **ACCEPTABILITY** despite being superbly "designed." One of **Raymond Loewy's** great successes was the Lucky Strike cigarette package. Sales boomed. **Loewy's** striking new package design brought increased employment for tobacco farmers, cut costs, raised prof-its, and looked superb. So cigarette design fulfilled his every professional promise. It also sharply boosted death rates for cigarette smokers, but in **Loewy's** period that fact was not under discussion. To scold **Loewy** for this would be like confronting the designer of a modern Sport Utility Vehicle and accusing him of melting Antarctica. It's the truth, but it isn't yet an **ACCEPTABLE** truth.

Loewy's greatest personal satisfaction as a designer came from working on Skylab, an orbital space laboratory. The Skylab habitat fell flaming out of the sky in 1979. Skylab proved to be a period objet and a technological cul-de-sac.

Skylab broke many of **Loewy's** self-set design rules: Skylab was never a saleable commodity, it bloated costs astronomically, it cost taxpayers a fortune and it had a mere handful of users. Skylab was also hazardous to its astronauts and people on the ground.

However, **Loewy** judged correctly that the awesome prestige of spaceflight outweighed other considerations. **Loewy** worked on spacecraft with particular devotion; when he appeared in a late publicity photo neatly kitted out in a NASA spacesuit, the old man never looked happier, or even, to judge by the grin on his photo, more at ease with himself and his self-created role in the world.

At the moment, humans have the thinnest of beachheads in settling space. The vast majority of today's orbiting objects are broken, obsolete junk. So **Raymond Loewy's** grand ambition to shape inhabited space stations looks rather archaic at the moment.

However: the clock will not stop ticking. Should the space-tides turn again for another set of argonauts, **Raymond Loewy** will look wondrously prescient. He will stand revealed as a prophet decades before his time, the Father of Spacefaring Interior Design. His glamorous mystique will only be enhanced by a fallow period.

9.

AN END-USER DRINKS GIZMO WINE

Let's return to my bottle of GIZMO wine, or, since some time has passed, let's open a fresh one.

It's easy for us to see that, in today's halting and new-fangled process of GIZMO wine, I (and you alike) are being methodically lured into practices that increase the market share of the manufacturer, distributor, and seller. And I'm the End-User: wine purchaser, consumer, unpaid promotional agent and incipient oenophile wine critic. Some of these are commercial practices; I pay. The others? They're technocultural practices. I do these things because that's the kind of hairpin I am.

But doing this pangs me in cognitive loads and opportunity costs. So how deeply can I, or should I, engage with this object, or any object? There are limits there, but not just limits of how much I can afford. They are limits of how much information I can process, and how effectively I can Wrangle that information into personal activity.

What's in it for me? What are my primary concerns about this object? How much it costs, of course—in a Consumer society, the price tag was always the stand-in metric for every other kind of measurement. But in a GIZMO society, mere price cannot be trusted. A price as low as literally free can mean the economic equivalent of a free kitten—I may get a free kitten, but then I have to deal with the consequences, with no exit strategy.

On eBay, it's now common to find objects offered for auction for a penny. I can have that object for a penny, because the point was to inveigle me into the auction process and a relationship with the auctioneer. If I'm given something for free, in a GIZMO-End User situation, then I need to be warily aware that this is almost certainly a loss leader of some kind meant to lure me into some tangled production chain.

What is the object doing to me? It's lightening my pocketbook—a serious concern. But there are a host of other unmet informational needs. Take that wine, for instance: I imbibed that material product and literally integrated it into my body tissues. What could be more personal than that? So what's inside this bottle? What the heck is it that substance I just drank?

I'm told the percentage of alcohol—for that is already legally required and expressed on the bottle label. That metric is probably roughly accurate. I'm also already cheerfully aware that alcohol is a major narcotic that can damage my brain and liver, that I'd better knock it off around heavy machinery, that I'd better not be pregnant, and various other well-meant interventions that have been either drilled into me by law enforcement, or clumsily attached to the product's packaging.

However—just as a kind of courtesy—I'd rather prefer to have instant, sophisticated access to a genuinely accurate ingredients list of this bottle's contents, down to the parts per billion. Why isn't this offered to me? Don't they know all that? If they don't know, why do I trust this bottle enough to drink the contents? And if they do know all that and won't tell me, why do I trust them? I know it wouldn't cost them much to spread that information around. Look at all the information they're pounding me with, already.

Who made my wine? Distant strangers, of course. I somehow imagine them to be cheery, suntanned Italian peasantry in the full healthful glow of EU agricultural regulations, but what if they're actually illegal African or Albanian immigrants? If that's the case, then I've been inveigled into oppressing these people under a veil of my own ignorance. That's certainly not something I would do voluntarily. Why do I collaborate with someone who forces me, through obscurantism, to do that against my will? I'd rather like a handy place to click where I can receive some kind of assurance that this was all on the up-and-up.

This bottle sure came a long way. How'd it get here to me? How much carbon dioxide got spewed into my planet's air in order to ship this object into my hands?

Now that I think about it, there must have been a jungle, a mountain range of externalities, currently obscured and invisible to me, that involved this object. That growing and fermenting of grapes... topsoil loss, tractor exhaust, chemical fertilizer, insecticide sprays, the fuels involved in heating and distilling all that liquid... I'm not supposed to worry my pretty head about any of that, but you know something? I know that I am paying for it somehow. Those phenomena do impinge on me; legal, social, ethical, environmental, all of them. They're not pretty, and neither am I. They should inform my decision about whether I buy that bottle and integrate its contents into my body.

What goes around, comes around. If I ignore distant consequences merely because they seem distant, then distant people will similarly inflict their consequences on me. That's a beggar-your-neighbor situation, a race to the bottom. But suppose I show them how the object came to be, and I link that information to the object. That would be "transparent production."

Is transparent production a good thing? Not entirely, but it's certainly a different thing, and one much better attuned to a society determined to thrive in the long term. How much of my own production should be made transparent? Well, that judgment depends on how often I get asked about it. If I find that I get asked all the time—there's a blizzard of queries on my Web site, let's say—then I might reveal it all in a rush. Of course, I lose some business confidentiality here. I throw myself open to the depredations of copycats. But I also throw myself open to people who might willingly participate and help me, even when they're not being paid. My best customers. Hobbyists, fanatics and devotees. Unpaid promoters and the cognoscenti. How can I design my production to daunt the first and encourage the second? Hmmmm.

This bottle arrived in my possession seemingly stripped of consequences, but those consequences exist. Where is this bottle going, once I empty it? The mythic moment of "getting rid of it," of throwing it "away," is supposed to be the sudden and total end of our mutual narrative as human and object. But that is by no means any end of any object. It's just the moment when I, the human, unilaterally decided to ignore the object. The object is merely semantically reclassified as "rubbish" and exported willy-nilly to the future.

Will that glass some day reappear, broken, under my feet? Under my children's feet? My grandchildren's? When exactly am I supposed to be no longer involved in this act of injury? Why did I touch this thing for a few moments and then doom it to an age in a landfill?

My relationship to this bottle of wine is a parable of my human relationship to all objects. That's a remarkably interesting field of investigation—though we think we know about it, it's a vast terra incognita, full of scary cognitive loading and crippling opportunity costs. Clearly some wise, sober personage—(not me, for I've been drinking)—should be investing some professional effort into clarifying the multiplex situations there. That vast wilderness will never be entirely clarified—that idea would be utopian—but it can be methodically explored and developed to a far greater extent than it is.

How? By whom?

My own single-handed effort is entirely unequal to that challenge. I simply can't know enough; the cognitive load is too great. I can do my part, I might Wrangle away in a vigorous fashion in some situations I know rather well, but I can't Wrangle all the world's technosocial issues all the time.

It follows this much of this activity should be done for me by other people. People with skill who really care about this subject. People well placed to guide me in these matters and help me out. A class of aware, well-informed, trained and educated people who can navigate their way through this field of complexity, negotiating the snaky processes of technosocial change and guiding them toward the sustainable. People who will make it their professional business, no, even their calling, their practice, their very mode of being—to create a human-object relationship that is as advanced as I can manage while still remaining acceptable to me. Who would that be, then?

Designers.

Who else is there?

With thanks and a tip of the fuel cell-powered hat to Scott Klinker, Cranbrook Academy of Art's 3-D designer in residence

10.

MEET THE SPIME

Scenario: You first encounter the SPIME while searching on a Web site, as a virtual image. This image is likely a glamorous publicity photo, but it is also deep-linked to the genuine, three-dimensional computer-designed engineering specifications of the object—engineering tolerances, material specifications, and so forth.

Until you express your desire for this object, it does not exist. You buy a SPIME with a credit card, which is to say, you legally guarantee that you want it. It therefore comes to be. Your account information is embedded in that transaction. The object is automatically integrated into your SPIME management inventory system. After the purchase, manufacture, and delivery of your SPIME, a link is established through customer relations management software, involving you in the further development of this object. This link, at a minimum, includes the full list of SPIME ingredients (basically, the object's material and energy flows), its unique ID code, its history of ownership, geographical tracking hardware and software to establish its position in space and time, various handy recipes for post-purchase customization, a public site for interaction and live views of the production change, and bluebook value. The SPIME is able to update itself in your database, and to inform you of required service calls, with appropriate links to service centers.

At the end of its lifespan the SPIME is deactivated, removed from your presence by specialists, entirely disassembled, and folded back into the manufacturing stream. The data it generated remains available for historical analysis by a wide variety of interested parties. That variety and those levels of interest are what you, a Spime Wrangler, consider of genuinely crucial interest. The SPIME is a set of relationships first and always, and an object now and then.

The key to the SPIME is identity. A SPIME is, by definition, the protagonist of a documented process. It is an historical entity with an accessible, precise trajectory through space and time.

A SPIME must therefore be a thing with a name. No name, no SPIME. This presents a serious semantic challenge. The labels that we attach to objects are never identical with the phenomenon itself; the map cannot be the territory. There is a frail, multiplex relationship between labels and materiality.

For instance, when I described that "bottle of wine" a while ago, everybody presumably knew that I meant a particular, coherent object. Yet that "bottle of wine" was a momentary congelation of material and energy flows. It has now become nameless, but it remains a process, still underway and mostly unknowable to me. That "bottle of wine" was once sunlight on Italian earth, lakes of grape juice, yeast in fermentation tanks, wood pulp for the label, colored inks, cork from Spain or maybe Portugal, plus a Californian grocery chain reacting to consumer trends and stocking a brand with some shelf appeal. Then I found it, bought it and consumed it. It continued as a dissociated flow of recyclable glass, consumed paper, hydrating fluids and a narcotic in my bloodstream, long since metabolized.

When I bought that "bottle of wine" I was also financing a situation that names and defines those complex flows as a "bottle of wine"—a technosocial set-up that allows me to interact with that object as a consumer item first and only, blindly uninvolved with its extensive history as pre-bottle and post-bottle. Buying and drinking it was my own business, and the rest of it is none of my business. How much of that business ought to be mine? Well—enough for me to have some reasonable security in the thought that my more general business won't come to a sudden, ugly, unsustainable end.

In an age of *artifacts*, I'm living off the land with most of my objects made by myself or my immediate kin. I know a lot about what I have, but I'm basically poor and ignorant.

In an age of *products*, I can engage in markets. But I'm just a gray flannel man in the crowd; I have to shut up and settle for what comes out of the assembly line.

In an age of GIZMOS, I'm an unpaid developer. I'm eyeballs, I'm keypunches, I'm Web site hits.

In an age of SPIMES, the object is no longer an object, but an instantiation. My consumption patterns are worth so much that they underwrite my acts of consumption. I can get PRODUCTS in profusion, but I've been kicked upstairs into management. I don't worry much about having things. I worry plenty about relating to them.

How? Mostly through naming. Naming enables the generation of pattern. Naming enables measurement. Naming gives me something to speak about.

In my relationship to objects, I have "advanced to the stage of science!"

"When you can measure what you are speaking about, and express it in numbers, you know something about it; but when you cannot measure it, when you cannot express it in numbers, your knowledge is of a meager and unsatisfactory kind: it may be the beginning of knowledge, but you have scarcely, in your thoughts, advanced to the stage of science."

So said Lord Kelvin. In an age of SPIMES, Lord Kelvin is not talking about physics. He's talking about the economy.

A MULTIPLEX, GLOBAL BUREAUCRACY
ALREADY EXISTS WHOSE PURPOSE
IS ATTACHING IDENTITIES TO OBJECTS. THAT IS
A NONPROFIT, QUASI-AUTONOMOUS

NON-GOVERNMENTAL ORGANIZATION KNOWN
AS THE UNIFORM CODE COUNCIL, INC.®
ALONG WITH ITS EUROPEAN TWIN, THE EAN
INTERNATIONAL, IT RUNS AN IDENTITY
REGIME THAT IS KNOWN AS THE GLOBAL EAN-UCC
SYSTEM – BETTER KNOWN TO THE PUBLIC
AT LARGE AS BARCODING.

The scope and scale of this enterprise is colossal. Barcoding has permeated commerce.

Having discarded my Italian wine bottle back in Tarzana, California, I'm currently sitting at a kitchen table in Belgrade, Serbia, where I gamely continue to labor on this book. There are twenty-three household objects sitting on this Balkan kitchen table. They are the common, quotidian objects that sit on this kitchen table most every day. There is nothing special about them, except that I just decided to subject them to an inventory.

Five of these everyday objects have barcodes, either adhering to them with gummed paper, or worked right into their surface finish. These five items would be two pens, the woolly winter hat, the packet of paper tissues, and the wine-bottle's local equivalent (which is a bottle of "Vuk Stefanovic Karadzic" brand Serbian plum brandy).

The phone handset on this table has its coding in another room, attached to the parent phone cradle. The phone cradle features two barcodes, a model number, and an ID number from the USA's regulatory Federal Communications

Commission, even though this phone is a machine in Serbia that has never been anywhere near America.

The TV remote control on the table is an extension of its extensively coded client, the television.

The stereo headset once had a barcode on its discarded packaging.

If you add the computer (which is no longer the laptop I was using in California, but an older, local model gamely crunching on a ported version of the same text), then we are immersed in identity coding. And this isn't the Los Angeles basin here, that sophisticated thicket of metropolitan consumerism—this is Belgrade, a city that is edgier in every sense.

I can also go trolling for kitchen-table objects that have Web sites embossed on them, inviting some End-User digital interaction. Then I get the plastic clamp, the brandy bottle, a pencil, and the blank compact disk (which sports five Web sites on its packaging alone).

Five of these objects: the saltshaker, the peppershaker, their stamped metal tray, and the wooden pencil holder— are Balkan heirlooms.

The coasters are too cheap to barcode.

The plastic cigarette lighter is so oddly and grimly anonymous that I'm pretty sure it was built in some Chinese basement and then filled with smuggled butane.

What we see in this household microcosm is a slow multi-decade, S-curve waves toward increased identity for objects.

Look at the variety here, as tomorrow composts today.
We have:

I.
Primeval *Artifacts*, handmade;

II.
Mass-produced PRODUCTS from the local Communist
era, pre-dating the local advent of identity coding;

II.
Trivial PRODUCTS too cheap or small to code;

III.
Coded PRODUCTS, including some strays whose codes
fell off or were dumped when they left the supply chain;

IV.
Two GIZMOS that are the remote adjunct interfaces
for a larger, fully-coded communication system;

V.
Coded PRODUCTS that also invite interaction
with a Web site;

VI.
A subset of Web site-only, non-barcoded PRODUCTS;

VII.

An awesomely complicated personal-computer GIZMO
whose <u>End-User</u> can Web surf with it, and go out to
briskly interfere at length with various supply chains,
potentially purchasing practically everything else on
this kitchen table through e-commerce;

VIII.

One radically GIZMOized PRODUCT with two barcodes,
(one glued on, one inscribed,) plus a Web site, an email
address, a complete postal mailing address, and a glue-
on, metal-and-plastic, interactive, electronic anti-theft
tag. That object would be the bottle of Serbian brandy,
which by this concerted effort has definitely established itself as the kingpin of Balkan consumerism.

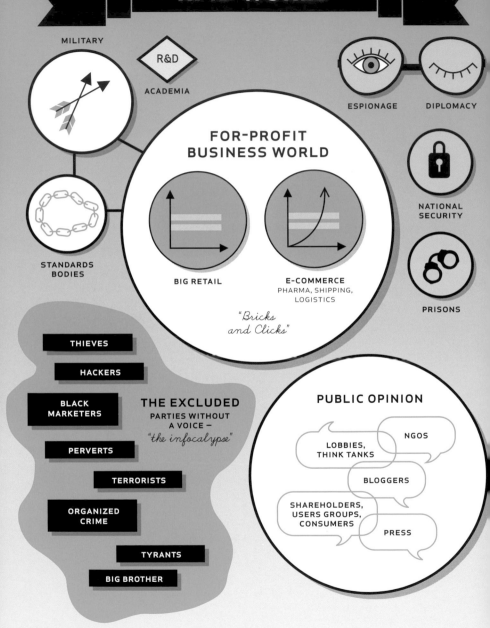

11.

ARPHIDS

The EAN-UCC revolution has been a colossal success. It's a coded delivery system, and it delivered what it was designed to deliver: by adding identity to objects, it enabled more accurate inventories, automated re-ordering, improved market analysis, a quicker movement of objects to and off the retail shelf, plus a sharp reduction in human errors in the supply and retail chain. And that achievement brought a lot of money to a lot of people.

Barcoding started as an R&D notion, then leapt the chasm of commercialization, into a firm foothold in the food & beverage business. It accelerated upslope in short order, into general merchandise, into healthcare, into government, and even onto the back of the book you are holding right now. Barcoding works. It is a great industrial advance. Pretty much any enterprise with a transportation chain can work more efficiently with barcoding.

On this very day, barcodes were scanned somewhere on this planet an estimated five billion times. The industrial payoff for exploiting barcodes has been 50 times larger in scale than was once estimated, when the system was first proposed, back in 1975. This success was also its bane, eventually. Now that people know about the full joy

FIGURE 4

and utility of a coded identity system for objects, paper barcodes are becoming obsolete.

The familiar system of black and white bars has passed the top of its S-curve. It is under threat from the new, radically disruptive, and far more capable EPC or "Electronic Product Code." Those aren't here yet. I don't have a single EPC object on this kitchen table. I know they are on the way, though.

Last night I watched the local television, and saw that the pet dogs of Belgrade were receiving injections of Radio Frequency ID identity chips. The local dog pound is being outfitted with an RFID reader, and when strays are collared, they'll be scanned. Then lost dogs do not have to have their homely pictures photocopied onto telephone poles. Lost dogs can be rescued quickly and returned to their grieving owners, which is sweet and nice.

But that's not the only way to describe what I just saw. We might also say that an RFID-injected elite of dogs will be returned to their owners posthaste, because these dogs now have a machine-readable identity. All other dogs are in grave and increasing danger. Belgrade is a rough town with a serious stray-dog problem. Being a Belgrade dog without an injected RFID may become a capital canine offense in relatively short order. We've got a yawning digital divide between the injected elite and the canine proletariat.

One could launch into a jeremiad at this point and point out that this grim dog-pound technology could be transferred at little cost and expense to, say, human vagrants, and then gypsies, ethnic minorities, political opponents, and/or anything else that moves, breathes or votes... but that doesn't much advance the analysis. What does advance the debate is the shocking realization that RFID chips are happening already *in Belgrade*. Serbian television news is promoting this technology to the general civil population as a public-service benefit. Who knows? This new coding system might even work as intended, at least in the sense of relieving some owners of worry—and bureaucratically liquidating some hazardous feral dogs.

Barcodes are made of paper. Electronic codes are electronic. That's why the EPC coded objects are coming; for the same reasons that electronics shoved paper aside in a host of other applications.

Paper codes are too slow, limited and small in scope for the ever-burgeoning needs and desires of the object identity enterprise. There's only so much data their one can cram into paper barcode digits.

A new-and-improved code would, obviously, import and store much more identity. It would also announce its identity more loudly, under a wider set of circumstances, to a wider set of scanning devices, and in more sophisticated ways.

Hence a new-model electronic identity: RFID, or "Radio-Frequency ID." RFID is busily composting EAN-UCC, even as we speak. The term "RFID" almost ranks with "EAN-UCC" in its acronymic ugliness. So henceforth, I will follow slang practice in the infant RFID industry and refer to radio-frequency ID labels as "arphids." We need to get used to thinking of these things as the seeds of SPIMEdom, not as some raw cluster of capital letters. We're better off referring to them with a neologism—"arphid"—that subtly implies some newfangled, infesting, autoreplicating plague.

First generation arphids barely work. They barely work in the following way: an RFID is a very small chip of silicon with a tiny radio antenna. An RFID tag can be as small as half-a-millimeter square and no thicker than a paper price tag. When it's hit by a blast of radio energy in the proper wavelength, the antenna will bend with the radio energy. The bending causes it to squeak a jolt of electrical energy through the attached silicon chip. The chip then automatically broadcasts a built-in ID code back through that tiny antenna.

That is a "passive" arphid, which already exist in large numbers. Passive arphids are cheap and easy to make in huge volumes. "Active" arphids have their own power supply, which allows them to get up to a wider variety of more sophisticated digital hijinks.

Arphids are tiny computers with tiny radios. They're also durable and cheap. It follows that one can build a new and startlingly comprehensive identity system with arphids. The arphid's antenna and chip get built into a weatherproofed, durable ID tag, to be glued, attached, or built-in to objects. A handy arphid wand (a "reader" or "transceiver-decoder"), beams radio energy into the arphids, then reads their unique codes as they bounce back out.

If a barcode is like a typewritten page of paper, then an arphid is like a written page on an Internet Web site. Those are both "writing" of a sort, but only a naïf could consider them the same. An electronically transformed means of production and distribution enables a wide variety of potent new behaviors.

Barcodes must be scanned within the visible sight of an optical reader. So barcodes require an attentive human reader focused on the paper code at hand. Arphids behave more like bats: their unique bouncing radar shrieks can be heard in total darkness, and while objects are in motion, and even all at one time, in massive arphid flocks. No deliberate human act is required to probe arphids with a radio pulse. An arphid-management system could be automated to inventory every arphid in its radio range, as often as you please.

For common, passive arphids, that radio range is quite short: less than ten meters. Since arphids are little radio stations, they have to behave that way through the laws of physics; as you move farther away from them, their coverage weakens and breaks up. This is considered a feature rather than a bug, because it prevents saturation of radio signals, a form of electromagnetic pollution.

Furthermore, metal and liquid—plumbing, wiring, metal appliances, a wide variety of everyday clutter—will reflect or absorb radio beams in the arphid wavelengths. This means that most real-world environments are full of radio shadows, where arphids become effectively invisible.

Otherwise, it would be an elementary matter to build a super-arphid reader inside some fiberglass van, and drive through urban streets trolling for rich people with a lot of arphid-tagged, purloinable stuff. Then thieves could rob the rich with maximal profit and minimal risk. This nightmare scenario is a little less likely to happen because arphids are so feeble from far away. Not that reading feeble signals is impossible to do. It's just expensive. Spy agencies like the NSA are sure to consider arphids of great interest, along with their little-known but long-abiding curiosity about the weak "Tempest" radiation that leaks out of computer monitors. Secretly snooping data from somebody else's arphids already has a name: it's a dirty trick known as "skimming."

So imagine: here you are, in tomorrow's emergent world of SPIMES, with your arphid tags, your arphid-reading wand, and some capable network nodes full of arphid-management software. Let's consider what can happen when you have the enabling means of a "mobile ad-hoc network." This means salting your arphids with a whole lot of arphid wands, placed every ten meters or so. These "wands" are not handheld scanning devices any more, so they might be better described as arphid "monitors."

A "monitor" should be cheap and easy to make, because it's basically just an active arphid. It's an arphid that

happens to have a steady source of power, a longer communication range, and a more sophisticated chip. It's
been moved from passive to active; it's now a boss arphid.
Monitors might be plugged into the wall, like contemporary appliances. Further into the future, they might be
wireless and running off an onboard micropower system.

The point of installing these monitors is that they can
communicate information about the arphids to one
another. Then they can filter that torrent of data and
move the valuable information over long ranges. They
become bosses, guards, co-ordinators. Add these monitors into the mix—active hubs of arphid data, repeaters,
relayers, linked to a global network
—and you have created an

INTERNET OF THINGS.

12.

AN INTERNET
OF THINGS

Given an INTERNET OF THINGS, you can read your arphids anywhere. Via Net, via cell phone, via satellite—it would seem that the sky's the limit.

But the sky's not the limit at all—for an Internet of Things, the sky is the *metric*. Global positioning satellites provide a splendid source of measurement for a space-time Spiming world.

Your arphid monitors are hooked into the satellite based Global Positioning System. Then your network become a mobile system of interlinked objects that are traceable across the planet's surface, from outer space, with one-meter accuracy, around the clock, from pole to pole.

A Global Positioning System is a literal world-beater—although satellite coverage breaks up whenever you move under a roof. A Local Positioning System, indoors, is handier yet. Global Positioning works by combining and analyzing signals from several cooperating satellites, up in space. The same thing can work on a local scale, inside a house.

If you have multiple monitors combined in a network, that means you can add arphid radio signals together, and triangulate them. It's an indoor, radar air-traffic control system for objects.

Real air traffic control systems are grim, complex bureaucracies, heavy with fail-safes. Who can make objects that integrate elegantly and dependably within an INTERNET OF THINGS? Who can make that system as relatively simple and inviting as, say, the Internet's Web browsers and Weblogs? It's a design space rife with profound opportunity.

You, a human being, don't want the cognitive burden of knowing what your host of objects is doing all the time. What you want is the executive briefing.

Management has its perks as well as its burdens. The drawback of becoming a Wrangler is a ceaseless struggle through changing fields of data and relationships. The benefit is that many previously knotty problems simply vaporize, they become trivial.

The primary advantage of an INTERNET OF THINGS is that I no longer inventory my possessions inside my own head. They're inventoried through an automagical inventory voodoo, work done far beneath my notice by a host of machines. I no longer bother to remember where I put things. Or where I found them. Or how much they cost. And so forth. I just ask. Then I am told with instant real-time accuracy.

I have an INTERNET OF THINGS with a search engine. So I no longer hunt anxiously for my missing shoes in the morning. I just Google them. As long as machines can crunch the complexities, their interfaces make my relationship to objects feel much simpler and more immediate.

I am at ease in materiality in a way that people never were before. Although I live in a much cleaner way than my forebears did, I am not achingly burdened by glum moral guilt about my acts of consumption. That's no longer a burdensome matter requiring constant conscientious decision-making on my own part. Instead, it's been designed into the metrics of the production stream. Whenever I shop, I shop with a wand in my hand. It would never occur to me to shop without a filter and an interface. And someone built that for me, it was designed—as a Wrangler, I need an interface for capitalism itself. In the old days, the best term for an idea like that was probably a "lifestyle magazine." Those toney, glossy little empires were the native haunts of the design profession. But those things were made of paper. They just sat there on a table. They couldn't *do anything*.

But now that design decisions are at my fingertips instead of stuck on paper, I can do a lot.

13.
THE MODEL
IS THE MESSAGE

Sometimes I really want an object, the thing *qua* thing, the literal entity itself, physically there at hand. At many other times, many crucial times of serious decision, I'm much better served with a representation of that object.

Suppose that I'm trying to create a new kind of object, to shape a new kind of thing. I don't want to be burdened with the weighty physicality of the old one. I want a virtual 3-D model of the new one, a weightless, conceptual, interactive model that I can rotate inside a screen, using 3-D design software.

Then I'm not troubled by its stubborn materiality; I am much freer to radically alter its form. I can see left, right, front, back, port and starboard. There's no gravity, no friction, no raw materials for making physical models. I'm spared the old exigencies of foamboard and modelling clay, of chickenwire frames and plaster.

I can change those immaterial plans as many times as I want. I can restore the changes, save the changes, erase the changes, export the changes. Because it's only data, it's weightless and immaterial. I can research vital information about it without lifting my hands from the keyboard or taking my eyes from the screen. I can show my work to a host of scattered co-workers at very little cost; I can offshore it to India, email it to China, get it back within the day... I've got an object processor! I'm crunching shapes! I'm processing objects! I'm no more likely to return to the older methods than authors are likely to return to typewriters.

After a while, once I'm used to this new routine, I don't even think of my model as "the model" any more. My model has become the central part of the creative effort. The modelling arena is where I shape my things. The physical object itself has become mere industrial output. The model is the manager's command-and-control platform. The object is merely hard copy.

In a SPIME world, the model *is* the entity, and everyone knows it.

Yesterday's old, creaky, limited 3-D modelling programs, such as ProE, FormZ, Catia, Rhino, Solidworks, are long-forgotten. Thanks to exponential, Moore's Law-style increases in processing, storage and bandwidth, an advanced SPIME 3-D modelling program can easily boast a finer grain of detail than the physical object it models. Instead of approximating form with a crudely nested set of polygons, a program with this capacity can generate more modelling polygons than the object in question has molecules. There's more stored in the map than there is in the territory.

Practically every object of consequence in a SPIME world has a 3-D model. Those that were not built with models have 3-D modelling thrust upon them. They are reverse-engineered: one aims a digital camera at the object and calculates its 3-D model by using photogrammetry.

While you're at it, you might as well photogrammetize your home and/or office, too. Your SPIME management software will surely become more efficient when it can measure and calculate the radio effects of the local walls, floors, ceilings, and furniture. Mind you, SPIME coverage is always patchy—always, because the laws of physics dictate that. No model is ever total and perfect. But you can always invest some more Wrangling ingenuity to make your Spiming just that little extra bit faster, more secure, less patchy.

How do you climb up that extra notch? With more processing speed, more storage and more bandwidth. How much does that cost? Something, but less all the time.

Where and when will you hit the SPIME limit to the measuring, labelling, and timing of made things, and this mapping of their environment? One might imagine (like Jorge Luis Borges in his prescient parable *Tlön, Uqbar, Orbis Tertius*), that the territory can't support the map. Sooner or later, reality will be historicized to the point of collapse. One is just bound to bog down and go broke in mud streams of sensor data, in ever-deeper sediments of bookkeeping.

Really, though? How, exactly? Why? For how long? Of course any particular processor, storage network or bandwidth network is subject to entropy and obsolescence. They will break, they will fail, they will have limits. But it may be that that process of deploying them, and extracting useful knowledge from analyzing that deployment, is endless.

Vannevar Bush said that science was the "endless frontier." Will we ever know so much about how things work that we can't afford to learn any more?

We can't know the answer to that. But we can surmise that a Wrangler, by nature, is someone pressing hard against these limits. So: having eagerly Wrangled my walls, floors and ceilings, and having contingently nailed down the balky behavior of my SPIMES, I now begin to wonder seriously about the other physical contents of this piece of space and time. Yes, to be sure, I have all my SPIMED objects named, coded, identified, and historicized—but what about their *environment*?

I am scandalized when it dawns on me that there are some "objects" in this area which are unnameable! Those would not be manmade objects at all, but environmental phenomena such as humidity...smog particles...pollen, magnetic fields, toxins, mice, dust mites, fluctuations in temperature... Certain local phenomena have not been subjected to a fully monitored historiography! Yet they can have measurable effects on both me and my precious SPIMES! Something must be done.

Here I take my technosocial cue from the experts of long-term object management, who are museum curators. Museum curators know well that the serious-minded care of precious objects over a long time must require both closely cataloged objects, and a closely monitored environment surrounding them.

Anything the museum curators of old used to do, I, as a modern Wrangler of SPIMES, can do at low cost and high intensity. So it's high time I added new functionality to my SPIME monitors. While the monitors are sitting there emitting and receiving those radio ID waves from identified objects, they might as well briskly measure light exposure, airborne pollution and pathogens, traveling microbes, pollen counts.... When inscribed into a silicon chip, functionality is very cheap. I've got bandwidth and storage galore, so why not add to my objects, a matter of course, a capacity to measure acceleration? Magnetic fields? Tilt? Chemical exposure? Any phenomenon that might trouble me and my possessions in any conceivable way? You never know when data like that might come in handy. After all, I don't have to think about it. I'll just explore it, store it, and maybe mine it later with some well-defined, handy interface.

Did I mention clocks? Of course every SPIME must have a clock, that sensor for time. *Shouldn't every object know what time it is? Fashionable items, perishable items— these goods have a time bomb ticking in them already! Anything with a sell-by date surely needs a clock! Given a long view, everything has a sell-by date. All things must pass; some of them just measure their way there.*

> IT MAY NOT SEEM THAT I "NEED"
> ALL THAT INFORMATION,
> BUT THAT'S AN OLD-FASHIONED WAY TO THINK.
> I DON'T "NEED" EVERY WEB PAGE
> ON THE INTERNET, EITHER.
> IT'S NOT A QUESTION OF DESIGNING
> AN INTERNET OF THINGS
> TO MEET MY SO-CALLED "NEEDS."
> IT'S VASTLY CHEAPER
> AND SIMPLER JUST TO ENABLE AUTOMATIC
> INFORMATION-GENERATING
> DEVICES AND PROCESSES, THEN SEARCH THEM
> MECHANICALLY AND CYBERNETICALLY,
> TO FIGURE OUT WHAT I "NEED."

I can't possibly waste my time trying to tell the Internet what's handy for me. That approach simply makes no sense. Just jam it all in there, all you folks everywhere! I'll make it my own business to winkle out what I need. You give what you give, and I'll give what I give. Then I'll search out my own answers in this blooming plethora. I can't waste time and energy telling you what I "need," or defining the problems of mine that you're supposed to "solve." I'll just use search engines to follow the tracks of other linkers and searchers. If it was good enough for people just like me, then it's probably good enough for me. It works for Google. I want a world that's auto-Googling.

Who owns the SPIME? This 3-D model awaiting its materiality.... This new-minted object on its way through a long set of human-object interactions? Who can alter it? What can they do with it? This ownership question in SPIME can never be settled. The fact that it's unsettleable is why there is money in it. There are no permanent solutions to SPIME questions. Only Customers and Consumers imagine that there are permanent solutions to physical ownership and intellectual property issues; End-Users know it's all a shell-game, while SPIME Wranglers don't even bother with the shell—they *are* the shell.

Wherever there is an insoluble intellectual-property question, there is a SPIME career. That's where I Wrangle. When and if it gets more or less figured out, I bump up the S-curve and I go Wrangle somewhere more advanced.

14.
FABBING

These Wrangling questions become especially acute with the advent of the "fabricator." We can define "fabricators" as a likely future development of the devices known today as "3-D printers" or "rapid prototypers."

The key to understanding the fabricator is that it radically shortens the transition from a 3-D model to a physical actuality. A fabricator in a SPIME world is a SPIME that makes physical things out of virtual plans, in an immediate, one-step process.

The fact that a fabricator is a wondrously cool notion doesn't mean that it's necessarily going to work in physical reality. Real fabricators would certainly be shot through with a wide variety of technical limitations, material constraints, shortcomings and holes. We can nevertheless confidently expect any SPIME technosociety to rejoice, agonize and sweat over fabricators, because fabricators are the SPIME equivalent of a Philosopher's Stone.

Shaping things, in one push-button step, from a virtual 3-D plan, is a staggeringly complicated manufacturing process. From the point of view of a SPIME Wrangler, however, it's a glorious, commonsensical event of well-nigh mystical simplicity. You just decide what you want to possess, push a button and bang! Lo, where there was once a 3-D schematic, there is now a newly minted object. You made a "fabject!" Build a SPIME tag into it, and it's ready to join the world!

The feedstock for a contemporary 3-D printer can be laser cured plastic, or heat-melted plastic dust, or liquid-sprayed starch, or glued sheets of cellulose, or, perhaps, some solid feedstock that was precisely chipped away. From a SPIME Wrangler's point of view, the ideal feedstock for a fabricator would be some renewable, recyclable, pollution-free goop whose material qualities—tensile strength, color, insulation, resistance to heat—are all specifiable on command. Materials like that don't yet exist. On the other hand, we've lacked a good reason to find them.

There can't possibly be just one such kind of universal cosmic wonder-blob fabricator food. That is a utopian notion. But the higher a SPIME technosociety climbs up that S-curve, the more rapidly it can compost all previous means of manufacturing.

As for the PRODUCT world, and the GIZMO world—when fabricators rule the earth, their systems get entirely dis-intermediated; all their tiresome nonsense of industrial assemblage and shipping gets raked up and baked down to compost.

A SPIME technosociety would want to route everything possible through the needle-eye of fabricators, in much the same feral way that a Customer society wants to lay its railroads all over the planet, or a GIZMO society lusts to put everything that matters to it: politics, business, news, gossip, jobs, sex, scandals, terrorism even—onto the Internet. There is a wild, irrational technosocial lust to achieve such things that no cost-benefit analysis can possibly tame. Economies will pulsate, groan and implode before an impetus of such profundity. If fabricators can be made to happen at all, they will be made to happen with gusto.

> We've reached a point where we need
> to take a breath now.

Let's try to summarize the central line of SPIME development. Identity is the key enabler.

1

First, we have the capacity for identity—
the code—which is modestly pasted onto the object.

2

In the second stage, a much thicker and more capable identity is embedded into the object, and that identity is historically traced.

3

In the third stage, the means of production
are re-engineered around the capacity for identity.
The object becomes an instantiation of identity.
It's named, and it broadcasts its name, then it can be
tracked. That's a SPIME.

WHY WOULD "IDENTITY"
EVER BECOME "MORE IMPORTANT"
THAN A REAL, NO-KIDDING PHYSICAL OBJECT?
HOW IS SUCH A THING EVEN POSSIBLE?

THE ANSWER IS FOUND IN A NEW MEANS
OF FOCUSING SOCIETY'S ATTENTION
AND ENABLING JOINT EFFORT.

Only a limited number of people can interact with any particular physical object. A real, physical thing is too small, too parochial, too limited to remain the center of importance for a large number of people. A real, physical thing occupies too small a piece of space and time. Most people in the world will never be able to see it or touch it. Its ability to interact with people is sharply limited. So only a limited number of people can contribute their skills and their insights to the process of that object's development.

For people outside that small circle, a physical object has limited importance. It might be a very important piece of technological development—it might be an atomic bomb, made in utter secrecy in a desert compound by a tiny elite of boffins. But most people would have no say in it. There is no way to make them care about it every day. They may be its victims, but they're not its stakeholders.

The object's virtual representations, however, can have stakeholders. For instance, it makes more sense to own shares of a company than it does to own physical pieces of a company. Like shares of stock, models of an object can be shown and distributed to a wide public. The models are more open than real objects. The models can attract a huge amount of creative effort worldwide—if they can find a method to cluster human attention.

Not everything in a SPIME world is a true-blue SPIME. Objects can also be *Artifacts*, **MACHINES**, PRODUCTS, even GIZMOS. But the SPIMES are the objects that are considered most important. SPIMES that get intensely Wrangled by many people can develop much faster than other objects. This means their S-curves are steeper. They have Rising Star quality, and they can return more on investment. They are interesting, glamorous, provocative. They are a locus of popular desire. People want to contribute to them, to know about the people who use them, to learn about them, to relate to them, to enter their fields. They are signifiers of power and desire.

They are the apotheosis of everything designers have been hired to do for the past ninety years.

15.

SPIME ECONOMICS

Those industries that can't or won't make the transition to SPIMES are in a dullard's line of work. Their S-curves will flatten. They will be occupied by *rentiers* and obscurantists. They will stagnate; like OPEC, they may have revenue, but they have no friends.

How is all this supposed to be made to pay, though? Well, the SPIME doesn't pay. The SPIME is economics itself.

Consider a bar of gold. It is precious. It has scarcity value. It's an *Artifact*.

But you can't compel anyone to do anything about your gold bar unless you hand it over. Your stake simply becomes his stake. Your gold bar can't appreciate in value; it has no return on investment; it does not harness human effort; it is a lump of metal.

You can advance to paper money, backed by some vague assurance by the authorities about gold bars, which are stacked up some place under their command and control. The presses whir, paper certificates appear in millions. That's a **MACHINE**. This paper money has a much higher traffic flow and is much easier fodder for the literate and numerate Customers.

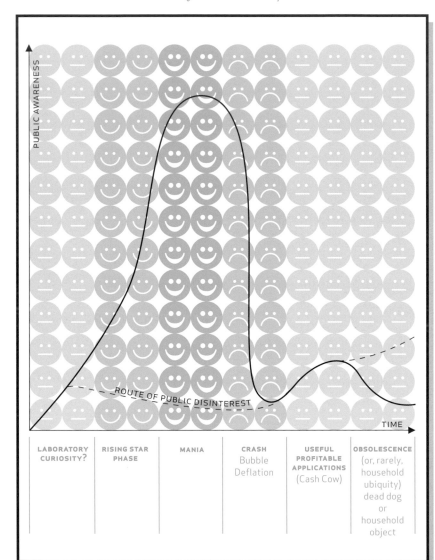

Cross a line of transition—cross the Line of No Return, cross the Line of Empire—and you can dispense with the gold backing. That money is worth money because the Consumer populace believes that it's fungible. It's backed up by the stark fact that it's consumed by everybody everywhere. Everybody knows about it, it's got a good brand, there's plenty and it works just fine. It's a PRODUCT.

Now dispose of the paper too, and metricize the global flow of electronic funds around the clock. Electronic financing certainly has its drawbacks and design flaws—it's profoundly unstable, it's fragile, it's always in flux, and it's subject to almighty panics. Currency transaction volumes are bigger than the worth that is generated by national economies. But electronic money girdles the earth seven times in a second. It's a GIZMO.

In a SPIME, value transmutes into a public interaction with past and future. It's not about the material object, but where it came from, where it is, how long it stays there, when it goes away, and what comes next. And just how long this can go on. Every market is a futures market.

Really? Yes. Consider your credit history. Your insurance. Your retirement funds. Would you rather have a bar of gold? How about a stack of paper cash? How long would you survive on that? The Line of No Return is already gone.

How about the Line of Empire? Has the world of GIZMOS crossed that line yet? Can it defend itself from attacks by nomads who refuse to buy into its logic and generate nothing it wants?

FIGURE 5

Read the newspaper. Look at your computer screen. You tell me.

Let's consider what electronic commerce looks like and how it differs in kind from earlier forms of economic behavior. Being an author, I do rather a lot of interaction with Amazon, an online bookseller that has become a generalized retail interface. I rather haunt Amazon. But I don't buy very much. I just involuntarily help them to compost the previous means of retail.

Let's consider what it is that Amazon, or any online bookseller, is up to when it sells what it claims are "books." When you buy books off the Amazon Web site, you do not touch any physical books—what you do is *perceive the virtual identity of books*. You never touch or see the physical book itself until it has been shipped to you through a physical distribution system.

A book listed on the Amazon site is much more than ink on paper. A book on Amazon bears the relationship to a normal paper book that an RFID tag bears to a paper barcode.

Once an Amazon book arrives in your physical possession, it looks, feels, and behaves like any ordinary book. Yet, in short order, you can use Amazon's data-mining capacities: you can find out its cost and its publisher, whether other editions have been published and the image of their covers; what other books that author has written; what readers think of the book and what other books those readers have bought; what other publications quote the book; and a host of even more intimate technosocial interactions.

You are heartily invited, even seduced at every opportunity, to contribute to this labor yourself. You can offer comments about the book, to be read by other Amazon End-Users. You can even sell the same book to other End-Users of Amazon, and Amazon, purportedly a book retailer, does not mind a bit when you usurp their industrial role and become a book retailer yourself, a direct "competitor," in earlier economic terms.

There are still many aspects missing from the spectrum of services provided by Amazon.com. It would be exceedingly useful and healthful to know the full composition of that book. How long will it last before yellowing and falling to bits from acid paper? What (possibly bioaccumulative) substances will subtly boil out of its glue and ink, settling into your body in years to come?

How much would it cost Amazon to add these interesting facts about the product they offer? Very little. Because somebody already knows—they're just not telling Amazon. Nobody's figured out that they could or should ask. Or that it might really matter to people.

Now imagine that we establish an Amazon.org, a social-software entity that hangs around the fringes of Amazon, answering these questions. Questions about objects. What questions? Not the profit-centric questions that obsess Amazon. The serious questions.

16.
THE DESIGNER'S QUESTIONS

Wim Gilles was a Dutch engineer, designer and design teacher. Back in the 1950s, Gilles decided to codify his method of analyzing industrial products, turning it into a useful algorithm for students. These questions are the questions a designer needs to ask when he plans to shape things.

Many parts of this 60-year-old **Wim Gilles** analysis still work just fine, while the aspects that have become different—well, those differences are excellent metrics for just how the relationship of humans to objects has changed.

Soon we'll get to **Gilles'** specific metrics, but first let's detour. Who might want to ask and answer these questions most frequently? Can we spread the labor around so that we can derive benefit without being crushed by cognitive loads and opportunity costs? Yes, because it is now entirely possible to ask these questions in gangs, on the Web, through social software, in "commons-based peer production." Open-source production of software is a maelstrom of Wrangling at the moment, because it's important. Open-source production of objects is an even larger challenge to the status quo.

So let's imagine that WE are a group of Wrangling enthusiasts, properly obsessed with our SPIMES. WE could be a government agency, a non-governmental organization (NGO), a group of hobbyists; WE could be a group of hobbyists forming an NGO and lobbying to sway governments, which is pretty much exactly what the Open Source movement is doing right now in Switzerland, Brazil and Spain.

But never mind the outcome of any particular incidental skirmish in the Wrangling. Every real industry is always surrounded by a huge technosocial haze of some kind: not just the paid employees, but regulators, educators, standards bodies, journalists, critics, advertisers, industrial trade groups, shows and expositions, labor unions, boards of directors, former employees, the retired, consultants, related industries up and down the supply chain, competitors, industry analysts, industrial spies, police investigators, fraudsters, forgers, fences, the invaluable people running the junkyards and doing industrial Superfund clean-up—you can name them. These onlookers outnumber people paid by the industry by orders of magnitude. If they can be united, the commercial enterprise they surround looks severely outnumbered and outgunned.

So let's see what this cluster of entities might do about turning themselves into a **Wim Gilles** SPIME COLOSSUS—

given that they have very capable and extremely cheap computational power, bandwidth, and data storage.

The first thing **Prof. Gilles** suggests is that we should assemble all the items with which our new-and-improved design might have to compete.

> "WHAT'S THE SCOPE AND SCALE
> OF THE INDUSTRY?"

Learning this was rather difficult in the pre-digital 1950s, but since WE at Amazon.org are sitting in the vast data shadows of Amazon.com and their ilk, that's dead easy for us. WE just contact our nutty completist-hobbyist friends, and set up a gorgeous Weblog database complete with photos and tech specs of everything made in the business. Unlike **Gilles**, WE don't stop with the living "competition." WE're interested in stuff that no longer exists, and the things that aren't yet made, and the things made in very distant countries with other markets, and the small-scale, odder things the industry used to make before they sold out and hit the big time.

So of course our database swiftly becomes far more comprehensive than Amazon's. Amazon merely wants to sell us available commercial things. WE, by contrast, want to know all about the works. And Amazon wants to promote them. WE don't need to promote them; we just want to know if they're any good.

Gilles now digs down into the basic characteristics of the object at hand.

> "WHEN WAS IT MADE?"

Every Spime Wrangler wants to know about "when." For the SPIME, it's all about timescales.

> "WHAT ARE THE
> FUNCTIONAL PRINCIPLES?"

Are WE engineers? No, WE can't all be. But some of us are. And the rest can help find the manuals. Then there's the "help desk." Google is already a much better help desk than most purported help desks supported by companies. WE're a vast user's group. WE can track all the most "frequently asked questions." WE may not know how these objects (supposedly) function, but WE know plenty about the endless struggles of people trying to make these objects perform in real life.

> "HOW DOES THIS OBJECT
> OBEY GOVERNMENT REGULATIONS,
> AND THE DICTATES
> OF STANDARDS COMMITTEES?"

Piece of cake to find that out. It's all public domain. With any kind of luck, in a SPIME world, government diktats are built right into the device specs. As for standards committees, they're commonly manned by greybeards, pundits, professors, retirees and minutiae freaks—pretty much exactly the kind of people that WE are, ourselves. A standards committee looks and acts a lot more like us than it's ever going to look like a retailer or a manufacturer.

> "WHAT DOES IT TAKE TO MAKE IT WORK?
> HOW MUCH ENERGY,
> HOW MANY RESOURCES?"

WE've got sensors. WE can measure all that for ourselves. And WE won't be much surprised if our real-world estimates differ wildly from the claims of the manufacturer and retailer. That's one good reason why potential buyers of these objects would want to consult with us, rather than them.

> "IS IT SAFE?"

Nobody ever knows what "safety" really means, but there is room aplenty for vivid public Wrangling in the turgid worlds of risk assessment and "Fear Uncertainty and Doubt." One thing is for sure—there is scarcely a commercial entity in the world with any spark of credibility when it comes to assessing the safety of its own products. As for government regulatory agencies, they are notoriously subject to "regulatory capture" by the wealthy commercial entities they supposedly govern.

But the capturers can't capture all the agencies all the time, and WE will make it our business to collate the output of agencies that aren't corrupted. Governments won't do that work of assessing their own regulatory performance, because governments are far too jealous of their own credibility—but WE will.

WE could get a lot of healthy public attention just by going to every national government and consumer-safety org in the world and summarizing their safety assessments, grouped under a single heading: the identity of the object in question. Then you put that on a handheld screen between the purchaser and the object on the shelf. Who's going to pay us to do all that work? That's likely not the shrewd question to ask. A shrewder question is: who's paying us not to?

> ## "WHAT'S ITS CAPACITY?
> HOW MUCH CAN IT DO?"

The manufacturers and sellers of a product are surprisingly unlikely to know this. That's because they designed and sold the object for a specific set of purposes that they themselves had in mind. WE know better, because WE are in intimate touch with the biggest otaku crank hot-rod fans of crazy post-consumer alteration, the cost-is-no-object fanatics who are using the object under circumstances never originally intended. You want to know what that thing can do when you strip it down, soup it up, and put it on the street? Ask us!

> "WHAT ABOUT HYGIENE?"

Hygiene is a subject never properly addressed under the previous technosocial regime. The 20th century's ignorance in this regard rivaled the 18th century's naivete about germs. So let's talk for a minute about, for instance, shoes. Did you ever notice that the soles of running shoes are made of heavy-duty, high-performance plastics? That's good, right? Because you want to run in those objects, repeatedly pounding them against hard surfaces with the full thudding weight of your body.

But when you do run (or even just knock around the house in your shoes and tracksuit, pretending to be athletic), what happens to the soles of those shoes? They wear away. They abrade from the burdens of your weight and surface friction. Their soles transmute into microscopic particles of high-performance plastic. And pal, you are *breathing* those particles. It's not as if you lit up that running-shoe like a cigarette and sat and smoked it—but, well, it *is* rather like that, actually. It's just *slower*.

You want to know where those particles go inside your body, and what that process does to you. It isn't pretty. But WE can explain that to you. WE took the trouble to find that out.

Through compiling data from hundreds of previous medical studies, WE are able to show strong correlations between various pollutants and asthma, testicular atrophy, cerebral palsy, kidney disease, heart disease, hypertension, diabetes, dermatitis, bronchitis, hyperactivity, deafness, sperm damage and Alzheimer's and Parkinson's diseases. WE can test our own blood for various pollutant loads and add that up on maps. Why not join in and have a look? You give a little, you learn a lot.

What can happen to you and your shoe-wearing body when a load of that byproduct builds up inside your system? Since inhaling abraded shoes was never before defined as a medical syndrome, WE and our org are the world leaders in exploring that phenomenon. You'll want to talk personally to the guys and gals who experienced that remarkable situation. They're in our "medical support group." Them, and their physicians and lawyers.

> "HOW DO YOU MAINTAIN IT
> AND SERVICE IT?"

Did you ever notice how many books there are for sale about popular objects, books like *The Missing Manual* or *The Repair Manual for the Compleat Idiot*? Did you ever wonder why companies are so bad about writing popular books about the objects they presumably know best? Well, there are three reasons why their books and manuals are lousy. First, all their public documents are vetted through a PR department, so they are basically promotional items. Second, they don't care much about you or what happens to you, after they take your money.

And third and most crucially, they don't know very much about their own stuff. Why? Because knowing about their stuff is not their reason for being. They are a commercial enterprise. So they are trying to be a lean, mean, businesslike business, making and maintaining these objects with as few paid employees as possible, so as to produce a high ROI for investors and a big output-of-value per employee.

Nobody pays commercial enterprises to fully understand what they're doing. That's not their metric.

WE, by contrast, know a lot about their objects. That gap in expertise is traditional. That gap is the only reason that designers ever existed. Designers know more about objects than the people who are making money from them. That's because designers aren't required to pay elaborate attention to shareholders and sales. Designers pay attention to things. They pay an intense, **Wim Gilles** style of attention.

> ## "HOW LONG DOES THE PRODUCT LAST?"

An absolutely critical issue for the SPIME Wrangler. If we were **Wim Gilles**, this would have been the first question WE asked! WE like to go right out to dumps, disinter dead examples of the product, document them with necrotic fascination, and put the images right on the Web site. WE're frankly fascinated by the ways in which they decay. WE can also shine 'em up, fix 'em, put 'em on eBay, and make a mint. WE certainly know a great deal more about *abandoned* objects than any commercial firm. Whenever a company dies, WE just subsume it.

> ## "WHAT ARE THE USES AND LIMITATIONS?'

"The Street Finds Its Own Uses for Things." And the Net—the Net is like all the streets at once, pouring their traffic together.

> ## "WHAT ABOUT PATENTS
> ## AND RIGHTS PROTECTIONS?"

Keeping up with intellectual property hassles is a full-time job. Most every complex object that comes off the chute is full of some kind of wicked barbed-wire snag, hidden in there by some MBA who gets it about "consumer lock-in."

But we're not Consumers. We're Wranglers.

"Shrink-wrap licenses." Who reads those?

"Disclaimer notices." Who bothers with those?

WE do. Because we're people who've already been cruelly nailed by shrink-wrap licenses and disclaimer notices. People come to us just because WE've numbered and counted all those mousetraps. WE Wrangle them for you.

You can help. The better they get at hurting you with all this surreptitious IP warfare, the more you need to talk to us.

> "WHAT ARE THE PRODUCT'S MATERIALS?"

Is if fabbable? If it's fabbable, then we're probably fabbing it already. If it's not fabbable, you'll want to talk to the guys who are willing to make it fabbable. You're going to love our SPIME Wrangler Fabbability guys. They are fanatics, visionaries, the very idea enchants them. They spend most of their time trying to make fabs that are fabbable.

"WHAT ARE THE METHODS
OF CONSTRUCTION?"

WE've got our guys who talk biomimicry, WE've got our guys who talk room-temperature auto-assembly, but... Okay, this may sound radical, but let's cut to the chase here. There's only one interesting, important method of construction: fabbing. If you still think otherwise, you want to talk to our guys who can fab stuff out of artificial diamond. *Diamond*, carbon atoms nano-assembled into diamond, right out of the white-hot vapor-deposition fab-spout. If diamond isn't durable enough for you, you're in the wrong universe.

"WHAT ABOUT PACKAGING?"

WE like trackable packaging in a network. If it's too dumb to know where it is and what time it is, WE don't even call that a "package."

Most of traditional packaging design was about the firm establishment of what they used to call "branding." WE Wranglers don't need to be told about "branding" by the paper surface of some package. What kind of lame customer-relations management is that? WE just wave a

SPIME wand at the package, and a SPIME management dashboard pops up on the handheld wand screen, linked to global databases like a mobile phone. Brand *that*, fella. If you're not in charge of what's happening there, it may well be that your worst competitor is.

What is a "brand"? It's a mark seared into the surface of something. Is that the best you can do in the way of establishing a relationship between us?

> "WHAT ABOUT STORAGE?"

WE're totally into inventory management. Inventory management is our very reason for being.

You know what the real story of storage is? Where's the place where manufactured objects spend the vast majority of their time on planet Earth? It's the dump. The junkyard. You never go there, but WE always do. WE'll be there waiting. The things tossed off the truck get torn to shreds and reverse-engineered.

> "WHAT IS THE EXISTING SALES PITCH?"

If you give the likes of us a "sales pitch," WE'll look at it like you offered us a hand-hammered flint rock. There's nothing you can tell us in a "sales pitch" that WE can't refute with a search engine in five seconds flat.

Whenever you coin a jingle, or trademark a slogan, and WE put those words into any Internet search engine, it's almost immediately going to lead us straight to your worst enemies. Those enemies won't be the first on the list—you'll be the first, because you spent so many millions making those words into popular taglines. But they're also a golden road to publicity for anyone who is particularly determined to hurt you. They can agglomerate the same traffic, that you built up at such cost.

> "WHAT ARE THE MEANS
> OF DISTRIBUTION?"

You distribute the fab data, then fab it on the spot. That's the Wrangler's favorite method, of course. There are some interesting distribution alternatives. For instance, you can leave a SPIME on the side of the road and let it offer ten bucks to any passer-by who can forward it toward New York. Your results may vary.

> "WHAT ABOUT PRICE AND VALUE?"

Now this is a truly fascinating topic. As SPIME Wranglers, WE're keenly aware that a deep engagement with identity can cause older pricing systems to crumble in unpredictable, nonlinear ways. For instance, imagine a world where every collectible appears in an issue of one.

> "WHAT ABOUT TRADE DISCOUNTS?"

WE *are* the trade! Who pays retail? Come on, WE're all insiders now.

> "WHAT IS THE VOLUME OF SALES?"

It's not about how many items jumped off the shelf this quarter. It's about how many objects there are in circulation, and what's being done with them. The volume of sales is trivial; it's developments within the installed base that tell us how many new ones may be needed or wanted.

> "WHAT IS THE IDENTITY
> OF THE PURCHASER?"

There are no purchasers. There are only Wranglers.

Who cares about "the purchaser"? If the purchaser's not in the Wrangling game, the purchaser is like a child. You want to know the identity of the early adapters, alpha geeks and stakeholders, on other words, all the people who most want to know about you. There are the people you want to know about, not the "purchasers." Get these people working in a direction you can leverage, and you can forget about mere "purchasers"—they'll show up as sure as lemmings pour into the sea.

> ## "WHAT'S THE IDENTITY OF THE USER?"

It's good that **Gilles** makes a distinction between "purchaser" and "user," but WE Wranglers would like to have some coherent ideas about the demographics of everyone who interacts with SPIMES in any way whatsoever. WE're not all that interested in pigeonholing people inside demographics—what interests us most is when people *transit across* demographics. A rural fundamentalist who somehow moves to a foreign country, triples her income and is now a refined international diplomat—she sounds like someone we might want to talk to.

> ## "WHAT DOES IT LOOK LIKE?"

Just as in the days of **Raymond Loewy**, it's still important to make a pleasing visual expression with a product. Being Wranglers, we want to know what the thing looks like at every stage of its lifecycle, not just when it's fresh from its shrink-wrap and styrofoam blocks.

"WHAT DOES IT FEEL LIKE?"

Yield to the hands-on imperative!

After asking these questions, **Gilles** tells us what to do with the answers:

> "Collate the positive and negative aspects
> of the products studied, and compare
> them in order to draw conclusions with which
> to formulate guidelines for the new
> product, which should possess as many of the
> positive characteristics as possible
> and as few of the negative ones as possible."

So that's it, right there. That's the crux of shaping a new thing that's rather like the older versions, except better. Not much to that, eh? Sounds like anybody could do it! Since I've now finished paraphrasing the work of famous designer **Wim Gilles**, I'll toss in another creative secret, for free: how to become a famous guitarist!

> "You put your fingers firmly on the fret board,
> and then move your other fingers up
> and down on the strings!"

So no, it's not that easy. Design is hard to do. Design is not art. But design has some of the requirements of art. The achievement of greatness in art or design requires passionate virtuosity. *VIRTUOSITY* means thorough mastery of craft. *PASSION* is required to focus human effort to a level that transcends the norm. Some guitarists have passion, especially young ones. Some have virtuosity, especially old ones. Some few have both at once, and during some mortal window of superb achievement, they are great guitarists.

The vast majority of people who play the guitar do it to amuse themselves and maybe few friends. These people are also the core of the audience for great guitarists, because, although they will never be great, they know what passionate virtuosity sounds like. They are cognoscenti, and without them, you may have genius, but you have no scene.

Then there are forms of music better handled by masses of people formally organized in orchestras. Or is that so? What if the principles of organization are being transformed? What I electronicize the sounds of musical instruments into sampled bits, combining that sonic product with new methods of assembly and distribution? Does that effort, make any sense at all? If it does, then how fast will that compost the old method? In what areas first, in what subcultures, in what applications? Where is the Line of No Return? Where is the Line of Empire?

PASSION and *VIRTUOSITY* don't vanish, but they may well manifest themselves in structures that were previously inconceivable. Until the 1920s, "industrial design" did not exist as a profession.

Let's imagine that an enterprise such as "Amazon.org" comes to exist. Is that enterprise going to "design" things?

I doubt that an org will ever win a design award. But it offers the potential to do what modern industrial designers always *talk* about doing, which is designing the industrial system itself. It's about re-shaping the great beast from start to finish. And over again. Over again. And over again. Making new mistakes. Learning from all the old ones.

Today's Net is a condition like the early days of the horseless carriage, where they used to ship them with a mockup of a wooden horse on the front, so that cars wouldn't panic the horses still in the street. It's in the sexy but vaguely absurd mode of **Raymond Loewy's** streamlined pencil-sharpener, a period artifact that was mocked by **Henry Dreyfuss**. **Loewy** had slipped a sleek, handsome monocoque shell over the pencil sharpener, but inside, both he and **Dreyfuss** knew that it still had the same old grinding mechanical guts.

The Web is a layer of veneer over 20th century industrialism. It's still a thin crispy layer, like landlord paint. It's a varnish on barbarism.

The heat is on. The varnish is cracking as the barbarism grows more obvious, harder to bear.

The 20th century's industrial infrastructure has run out of time. It can't go on; it's antiquated, dangerous and not sustainable. It's based on a finite amount of ice in our ice caps, of air in our atmosphere, of free room for highways and transmission lines, of room in the dumps, and of combustible filth underground. This is a gathering crisis gloomily manifesting itself in the realm of bad weather and resource warfare. It is the legacy we received from world-shaping industrial titans such as Thomas Edison, and Henry Ford, and John D. Rockefeller— basically, the three 20th century guys who got us into the Greenhouse Effect.

It's no use our starting from the top by ideologically re-educating the Consumer to become some bizarre kind of rigid, hairshirt Green. This means returning to the benighted status of Farmers with *Artifacts*. End-Users will always legally and politically evade any effort to reduce them to the status of Consumers, and even Consumers will stoutly refuse to become Customers or Farmers; they know that any such effort of repression is the path of the Khmer Rouge and the Taliban.

THE ONLY SANE WAY OUT OF A TECHNOSOCIETY
IS THROUGH IT, INTO A NEWER ONE
THAT KNOWS EVERYTHING THE OLDER ONE KNEW,
AND KNOWS ENOUGH NEW THINGS
TO DAZZLE AND DOMINATE THE DENIZENS
OF THE OLDER ORDER. THAT MEANS REVOLUTIONIZ-
ING THE INTERPLAY OF HUMAN AND OBJECT.
IT MEANS BRINGING MORE ATTENTION
AND ANALYSIS TO BEAR ON OBJECTS
THAN THEY HAVE UNDERGONE.
IT ALSO MEANS ENGAGING
WITH THE HUMAN BODY AND ITS AFFORDANCES,
WITH OUR HEALTH AND OUR EASE
AND OUR COMFORT,
WITH OUR WORKING ENVIRONMENT,
OUR HOME ENVIRONMENT,
WITH OUR LUNGS, AND OUR SKIN,
AND OUR BONES.

17.

TOMORROW'S TOMORROW

Look hard at the people who use the Internet most often. You'd think these <u>End-Users</u> would be pretty far removed from the grim exigencies of manual labor; after all, it's not like they are coal miners.

We don't need to wax all stereotypical here; doubtless there are coalminers working today whose creamy skin is spotless and whose hair is a crisp bouffant. But, well, hang out with real hackers, sometime. I do that. I do a whole lot of it, because they are interesting. These masters of the digital universe, the heavy-duty programmers who build and maintain the Internet, they are commonly portly guys with wrist supports, thick glasses and midlife heart attacks.

They weren't born that way. They didn't get that way by accident, either. They got that way by chronic, repeated abuse. That's not a digital problem, that's a physical problem. It's still about an industrial system that cruelly sacrifices human flesh for the sake of dysfunctional machinery. They sit, type and stare in screens. All day, every day. It ends up hurting them. It hurts them in ways that are slow enough and subtle enough to steal up on them.

The step after the SPIME Wrangler—tomorrow's tomorrow—is neither an object nor a person. <u>It's a **Biot**, which we can define as an entity which is both object and person</u>.

A **Biot** would be the logical intermeshing, the blurring of the boundary between <u>Wrangler</u> and SPIME. This is happening now, but we can't perceive and measure it.

Today, every human being, everything that breathes, carries a load of industrial effluent. The industrial and natural worlds have interacted long enough and powerfully enough to become a kind of planetary froth. Trees and grass have been absorbing smokestack spew for two centuries; detritus, fertilizers and pesticides have washed off the continents and been thoroughly churned into the seas. The human body breathes, eats, drinks, excretes, assembling flows of material and energy, and since a human body lives in a froth of microscopic rubbish, people are increasingly composed of effluent.

A human body can be understood as a sponge of warm saltwater within a shell of skin; so everything we emit ends up partially within ourselves.

Some artificial substances are "bioaccumulative;" our metabolisms preferentially suck them out of the biosphere and try to make structure out of them. These processes are involuntary and take place beneath our awareness.

A **Biot** is somebody who knows about this and can deal with the consequences. He's in a position to micromanage and design the processes that shape his own anatomy. The techniques that will allow individuals and groups to do this cheaply, effectively and as a matter of course are several decades away.

How far away? Let's hazard a guess.

If the Consumer/*PRODUCT* epoch lasted from World War II to 1989, and the End-User/GIZMO epoch from 1989 till 2030 or so (another forty years), and if the Wrangler/SPIME epoch managed about the same time span, then the advent of the **Biot** would arrive around the year 2070. I would guess that 2070 is a reasonable date for a situation in which human biochemistry is well enough understood to become a medical-industrial complex. In a **Biot** world, the leading industries are not *Artifacts*, **MACHINES**, *PRODUCTS*, GIZMOS or SPIMES, but technologies for shaping human beings. The people who do this are both the shaper and the thing shaped, the user and the tool in one.

The driving technologies of a **Biot** technosociety would be cybernetics, biotechnology, and cognition. We're well into the first one, struggling to make sense of the second, while the third remains an unknown world.

The future combination of the cybernetics and biotechnology suggest a technosociety where objects are fabricated by biological methods—not *grown*, necessarily, but produced through redesigning and exploiting the biochemical processes by which living tissues grow. Living tissues have many industrial advantages: They grow at room temperature, they can use solar power, their products and effluents are mostly compostable, they scale rather easily, and, basically, we human beings get built that way ourselves. This means that anything we learn about that industry can handily double as "medicine."

However, living tissue grows slowly by the standards of mechanical industry. We humans could always sharpen a new rock a lot faster than a tiger could grow a new fang.

If you want fast, bulk industrial action in biochemistry, you have to go microscopic. Producing industrial material, through biological methods, in bulk, means using microbes. Microbes can double their volume and number in minutes. Microbes gave us oil, gas, and the White Cliffs of Dover. Plus, microbes live on us. Human beings are a vast playground for microbes. Their misbehavior kills us in large numbers. We need to understand them in order to better understand ourselves.

Humans begin as microbes. If you want to tell our own history the way we would tell the history of a SPIME, then we begin as fertilized eggs, not as babies with a name. Birth is when we get our identity. There are nine months of work before birth.

For a **Biot**, manufacturing and metabolism are the same field of study. Understanding metabolism means understanding the action of hormones, neurochemicals, DNA, agencies operating at microscopic scale with raw materials present inside as at a few parts per billion. This implies an ability to track and trace that is vast indeed. A Biot would trace the history of traces.

The ultimate consumer item is the Consumer. There is no metahistory we find more utterly compelling than our personal metahistory. The world has many forms of reward and gratification, but being alive and healthy underwrites all the rest of them.

A **Biot** would understand and manage the living processes going on within herself. If she can really do this, then the previous human technosocieties compost quickly; the rules have changed as never before, for age-old limits of the human condition have been overcome. Limits are crossed; the Line of No Return, the Line of Empire.

As a **Biot** you cannot go back, for you had already outlived any human lifespan; you cannot be overthrown by the previous order, because your new capacities are simply too great for reactionaries to combat. You are no longer human. Not that this lets you off the hook in any way; you have a wide variety of interesting, challenging problems. It's just that none of those are human ones.

The human condition isn't abolished overnight, it isn't obliterated. It is composted. If you're not afraid to watch such things, but if you sit and watch with patience and an inquiring spirit, knowing what you are looking for, then you will see them melt away into air.

18.
UBLOPIA OR OTIVION

Visionary futurists have a remarkable quirk. They tend to enforce the gravity of their prophecies by asserting that they will come true—or else. The stakes could not be any higher (so they will tell you). It's Utopia, or Oblivion—*my way to futurity, or the handbasket to hell!*

I frankly care nothing for "Utopia" or "Oblivion." If my long romance with futurism has taught me anything, it's that neither of these terms has any meaning. They are mere verbal gasps of intellectual exhaustion. They mean only that the futurist has exhausted his personal ability to confront the passage of time.

Either everything is arranged in a permanent system of which he approves—that's "Utopia." Or else every event that might be of possible interest to him can no longer take place, which would be "Oblivion."

These two archeologisms, "Utopia" and "Oblivion," are definitely showing their age, and, like the wacky shibboleths of some ancient theology, they are getting in the way of our ability to creatively affect the course of future events.

No society is ever going to achieve perfection through an ideal technosocial set-up that achieves its every wish.

We can't make everyone happy and contented. We couldn't make one single person happy in a Utopian sense, even if we devoted the entire productive capacity of everyone alive to the task of giving that single person a Utopian experience. It can't be done.

People can't outguess themselves through planning. Their needs, and desires, and wishes defy prediction, for they are hierarchical, nonlinear, time-bound and inherently conflicted.

Hierarchical: People require air, water, food, sleep, sex, safety, stability, consistency, social acceptance, camaraderie, attention, recognition, admiration, self-esteem, self-actualization... Kick one of the lower rungs out, and the ones above it all collapse at once, they become irrelevant. People's needs aren't a checklist or a restaurant menu, they can't be outguessed.

Nonlinear: if you need sleep, you don't need sleep a little, then twice as badly, then three times; your need to sleep soars uncontrollably, becoming non-negotiable and absolute. Once you've had enough sleep, you need to wake up; you can't stay asleep even by a determined act of will. Human experience is full of wild needs like this, unpredictable urges that spring up like hurricanes from a butterfly's sneeze. The world right now has millions of perfectly normal, rational, well-meaning fellow citizens who are absolutely dying for a kiss.

Time-bound. Utopia for a ten-year-old would never involve puberty; she'd never think to ask for such a radical transformation, and if fully informed about the trouble such a change was sure to cause her, she would surely be appalled.

Inherently conflicted. People aren't perfectly autonomous like polished ball bearings. They want to participate in a love relationship, a family, a clan, a neighborhood, a city, county, state, nation and planet, but the interests of these entities don't coincide. We can't be all things to all people all the time.

A successful human lifespan involves a lifelong maneuver through a transmuting landscape of hierarchical, nonlinear, time-bound and inherently conflicted demands. It's a homeostatic tumbling with enough flexibility to allow effective action, but enough continuity to avert terrifying chaos. Oddly, we don't have a term for a person or a society that excels at living in this sensible, everyday way.

Maybe we could call that

"Ublopia."

Then there is that troublesome term "Oblivion." What could it mean? Let's assume the Sun explodes and reduces the Earth to a cinder, annihilating all known life. That's clearly a very severe situation, and it's not blatantly impossible; for all we know, it could happen tomorrow morning. Suppose it did. That would create oblivion. But who would make a fuss about that? We'd be in the situation of the tree that fell in the forest with no one to hear.

"Oblivion" is un-regrettable. We can dread it in advance, but once it's done, there's nothing left to be concerned about.

What we really ought to fear is not "Oblivion" but irretrievable decline. This would be a grim situation in which we all knew that humanity's best days were behind us, and that none of our efforts, however brilliant or sincere, could redress the mistakes humankind had already committed. Hope has died within us as a species; our hearts are broken; animal vitality keeps us on our feet, but the only satisfaction we have lies in inflicting harm on ourselves and others. Despised by ourselves, we are an active source of evil to others.

This isn't "Oblivion" but a genuine, rather common way of life; visit prisons and mental asylums, and you'll see that it's as real as concrete. This misery is so comprehensive, painful and repugnant that it ought to motivate us far more powerfully than mere oblivion ever could.

We might call that desolate state of mutilation,

"Otivion,"

assuming that we felt that we had to name it, somehow.

We need to understand what that threat is: the knowledge that tomorrow will be like today, only certainly worse. Because we ourselves are worse: we're collaborators in our own corruption. We need to understand that we really don't want to find ourselves in a world that fits that description.

And in order to avoid that fate, we need to work. We need to tear into the world of artifice in the way that our ancestors tore into the natural world. We need to rip root and branch into the previous industrial base and re-invent it, re-build it. While we have the good fortune to be living, we should invent and apply ways of life that expand the options of our descendants rather than causing irreparable damage to their heritage.

The technologies in most critical need of reform are the biggest ones. These are the ones that have spread themselves throughout the technosocial fabric, into commerce, infrastructure, governance and culture. These are the technologies that throw the most weight around. They're not the fanciest ones, but the common, simple ones that most everybody is used to doing.

We're living at the tag-end of the Information Age. The Information Age is the successor to the Space Age, which was preceded by the Atomic Age, the Jet Age, the Radio Age, the Aviation Age... The consequences of those so-called "ages" are still here now. We just composted them under new levels of technological novelty. If we stopped inventing new technologies today, that installed base of older technologies would continue to transform us and our environment. And we would likely die of those changes. Because those technologies are not sustainable.

We need to understand technology with a depth of maturity that mankind has never shown before. We need to stop fussing over mere decade-long "Ages" and realize that there are only three basic kinds of "technology" truly worthy of civilized use. None of them are entirely possible as yet.

1

The first kind, and likely the most sensible one, is technology that can eventually rot and go away all by itself. It's materials and processes are biodegradable, so it's an auto-recycling technology. The natural environment can do this kind of work for itself, while also producing complicated forests, grasslands and coral reefs, so, someday, an artificial environment ought to be able to biomimetically mimic that achievement. This doesn't mean merely using available "natural materials" that we repurpose from crops or wilderness. It means room-temperature industrial assembly without toxins. We're many decades away, maybe centuries away, from mastering feats of that order. It's no use sitting on our hands about it, but it's too distant a prospect to be of immediate use.

2

The second kind of technology is monumental. These are artifacts deliberately built to outlast the passage of time. This is very hard to do and much over-estimated. Many objects we consider timeless monuments, such as the Great Pyramid and the Roman Colosseum, are in fact ruins. They no longer serve their original purposes: a royal tomb and a giant urban playground, and they no longer look remotely like they did when their original builders finally dusted off their hands and demanded their pay. But at least these "monuments" don't crumble unpredictably, leach into the water table and emit carcinogens while they offgas. Objects that do that kind of mischief, and lo they are horribly many, should not be properly considered "technology" at all—they are a Ponzi scheme that exports its costs to our descendants.

I doubt we'll ever build a host of useful objects that are also multi-generational heirlooms. They wouldn't burden the environment much, but they wouldn't meet our needs, because our needs are always changing.

3

The last kind of decent technology is the kind I have tried to haltingly describe here. It's a fully documented, trackable, searchable technology. This whirring, ultra-buzzy technology can keep track of all its moving parts and, when its time inevitably comes, it would have the grace and power to turn itself in at the gates of the junkyard and suffer itself to be mindfully pulled apart. It's a toybox for inventive, meddlesome humankind that can put its own toys neatly and safely away. That's a visionary idea. It may not be possible, but at least the concept is new. Nobody ever thought in quite that way before.

But it's not enough to think about that, or even write about. If it's to be of any use to humankind, it will have to get done.

I hope that you're the kind of person
who can do it.

endtroduction

I was mulling over Shakespeare's observation that the future is an "undiscovered country." No, that's not true; I was watching late night cable and stumbled across one of those forgettable *Star Trek* films from the 1990s, with that phrase in the title. But then I remembered that Shakespeare wasn't referring to the future, he was referring to death. Actually, that's not true either; I looked up the phrase and my search engine returned the proper context from *Hamlet* (which, yes, I really did read, but so long ago that it's an unrecovered country). This mix of the high and the low, the dread and the absurd, constitutes the future, and that's what this Mediawork pamphlet is about. How we might better shape our futurity has long been the province of Bruce Sterling.

Best known for his eight science fiction novels, Bruce also writes short stories, book reviews, design criticism, opinion columns, and introductions for books ranging from Ernst Jünger to Jules Verne. His nonfiction works include *The Hacker Crackdown* and *Tomorrow Now*. He runs the Viridian list on environmental activism and postindustrial design. He is a longtime contributing editor for *Wired*, for whom he also writes the "Beyond the Beyond" weblog. Bruce is a polymath with a love of language, an activist with the mesmerizing presence of a street preacher, a futurist who really understands the past and present, and to top it all off, an absolute fiend for design.

Just after he finished the final draft of this book, Bruce moved to Pasadena for a year to become Art Center College of Design's first ever "Visionary-in-Residence." No one was quite sure what the position required, besides moving into the Media Design Program's studio. But Bruce understood that the first thing he needed was a logo, so he immediately organized a Viridian on-line competition for a Visionary-in-Residence tee shirt. Bruce came to Art Center to live the precepts he lays out in this book, working with the next generation of designers to address one of my own concerns: that for some time—since the end of the space race, maybe—our culture's been running a vision deficit. We haven't been able to see the future's forests through the present's trees. If there's one thing we ought to be able to do, it's train a new generation of visionaries: people who not only can imagine a better future, but can visualize and design it – in other words, shape the things that constitute our made world.

There's no one better suited than Bruce Sterling to take a vision deficit and turn it into a surplus. In this volume, he's aided and abetted by the renowned Lorraine Wild. Lorraine (along with her colleagues at Green Dragon Office) practices design in collaboration with architects, artists, museums and publishers. A professor at the California Institute of the Arts, she publishes her design criticism and history widely, most recently joining the editorial board of designobserver.com. She is a partner of Greybull Press, a Los Angeles-based publisher of photography books, and a principal in the new design supergroup, Wild LuV. She's a winner of the Chrysler Design Award for Innovation, was named one of I.D. Magazine's "I.D. 40," and has been honored by the Smithsonian/Cooper-Hewitt National Design Museum, the National Design Awards, the New York Art Director's Club, and the AIGA, among others.

Bruce and Lorraine are friends, having built a mutual admiration society at Mike and Kathy McCoy's yearly High Ground design conversations in Colorado. In California, I got Bruce and Lorraine together to shape *Shaping Things*, which we often did over sushi. Perhaps that's the culinary explanation for why – in this resolutely rational book – there's a hint of wabi-sabi, the transcendent impermanence so beloved of Japanese art and craft. Lorraine was able to go off with the raw text and come back with this example of what I term radically classical design. She's an experimentalist who values readability, a deployer of discourse who seduces with form. The thing she's shaped here has resonance.

I'd like to close by offering my sincere appreciation to the two great champions of this series, Doug Sery at the MIT Press, and Joan Shigekawa at the Rockefeller Foundation. Finally, I wanted to thank all of the people who contributed to Mediawork project, the pamphlets, the WebTakes and the Web Supplements. For *Utopian Entrepreneur* – Brenda Laurel, Denise Gonzales-Crisp, and Scott McCloud; for *Writing Machines* – N. Katherine Hayles, Anne Burdick, Erik Loyer, and Sean Donahue; for *Rhythm Science* – Paul D. Miller aka Dj Spooky that Subliminal Kid, Cornelia Blatter and Marcel Hermanns (COMA Amsterdam/New York), Peter Halley, and Casey Reas; for *Shaping Things* – Bruce Sterling, Lorraine Wild, Stuart Smith (Green Dragon Office), John Thackara, Anne Pascual and Marcus Hauer (Schoenerwissen/Office for Computational Design). You can find out more and interact with their work at <mitpress.mit.edu/mediawork>. See you in the future.

Peter Lunenfeld
EDITORIAL DIRECTOR

designer's notes

For years now I have been attending a gathering of designers at the somewhat inconveniently located Rocky Mountain compound of my old teachers, Mike and Kathy McCoy. The McCoys invite a multi-disciplinary group for two and a half days of what only can be described as "design chat," though I guess I am being a little coy here as to the revitalizing, if not downright necessary nature of the discussion that ensues: which is why year after year we get on trains, planes and automobiles to make our pilgrimage to the event called "High Ground." During the weekend, attendees give 10 minute talks on what that they are thinking about, or working on, which the group then hashes over. At the end of the weekend, we analyze our discussions as a way of taking the pulse of design at that moment. Bruce Sterling has a been a High Ground regular for a while, always giving the rest of us the distinct impression that he was there for anthropological observation, to record the folk habits and foibles of our tribe, fodder for his sci-fi-design fantasies. But he participates, too, and in the summer of 2004, he delivered a rant about the future, about the absolute necessity for re-thinking the way we do just about everything, if the earth is to survive, if future generations will not loathe us. It was one of the more moving statements ever delivered up there in the thin air of Buena Vista.

His High Ground diatribe was, I quickly realized, the draft for this book. It was there that his analysis of the techno-social ecology of objects and call for the development of spimes made their début. And while there was an urgency to his message, a challenge, and plenty of black humor (for what else can we do in our current predicament but laugh), there was optimism at its core. Sterling respects design too much to imagine that it has no answers. So I felt that the design of this text would have to be driven by the same attitudes expressed by the author, which were so humanistic and so heartfelt. This book had to contain a riposte to the reigning visual clichés of technology: the intimidating complexity, mirrored surface of omniscient virtuality, and corroded surface of digitalization run amok.

During that same summer, two teenagers (one niece, one cousin) were house and studio guests of mine for several weeks. They hung around scanning stuff for me, taking pictures, IM-ing incessantly, posting on their blogs, and silk-screening t-shirts on the lawn behind the house. I thought how the future Bruce spoke about so urgently was theirs, and noticed that they are not freaking out about anything quite yet. However, there is an urgency and an agency in the way that they connect with the world that felt right. Their attitudes were not naïve but instead unfettered by the consciousness of critique to clip their wings (and powered by a deftness with technology that we all know is taken for granted). The personal identity and expression visible in everything they produced, from the cut-and-paste websites to science fair graphics (on the genome, for gawdsake!) were inspirational to me. I wanted to channel that energy, that incubation of the future, and bring it to *Shaping Things*.

No time-machine will turn me into a teenager, but I am a perennial student. It is the language of the hi-lighter and the written-over text-book, the embrace of new information that I see in my nieces's mark-making, the enthusiasm my daughter brings to learning to read, and my own studiousness that I have tried to infuse into this book (with the invaluable and steadfast assistance of Stuart Smith). I have had to read it about ten times in order to design it, and *Shaping Things* has not gotten old yet. Mr. Sterling (has no one given him a PhD yet?) has a way with words, and I was most anxious to shape this book in a way that would enlarge his already expansive view of the promise of design and the future, so that it can be seen in more ways than one.

As I worked on the design and mulled over Bruce's typology of objects, it dawned on me that the book as a communicative device is one of those rare things that actually is, to use his terms, an artifact, a machine, a product, a gizmo, and gosh-darn it, a spime as well (certainly the barcode is the gateway to that condition) all at the same time. So, while I accept his statement that "tomorrow composts today" and all futility that that implies, I do think that the book, the object upon which I have lavished so much thought and attention, is somehow an exception in the techno-social ecology he describes. Which is why, in the end, the savvy Peter Lunenfeld, upon commissioning such an interesting, forward looking set of texts he edits for this Mediawork series, still would have to orchestrate the process to its logical conclusion: as pages printed on re-cycled pulp with ink, glued together, wrapped and boxed and shipped and unpacked and displayed (or maybe not) until it reached the hands and eyes of you, dear reader, in a format not too different than that devised by the various Plantins of Antwerp who, in the sixteenth century decided that it would be nice to have books small enough yet important enough to carry in one's pocket.

No matter how far-reaching the ideas, in other words, we are always back to the reliance on a mulch of all-too-physical stuff, only here it has a life-span that way outlasts our own organic selves... which somehow reminds me of an inscription I saw in Houston, Texas year ago, on a tow-truck painted solid black:

"Al's Auto Mortician /
Sooner or later, I'll come and get ya..."

My work on this book dedicated to Rosalie Wild, Sarah Wild, Ana Xiao-Fei Wild Kaliski, Bruce Sterling's daughters, Peter Lunenfeld's kids, and all the other young ones bound to live in the world we are busy "fabbing" everyday.

Lorraine Wild

author's acknowledgements

As an author and journalist, I owe debts
to the design world for a ceaseless
flow of stimulating ideas and great copy.
This is my opportunity to thank a few
people in particular.

First and foremost, Mike McCoy and
Katherine McCoy of the High Ground
Design Conversation. The McCoy design
salon in the stellar, clarifying heights
of the Rocky Mountains is the most consis-
tently interesting meeting of minds that
I've ever attended. I've never yet left
High Ground without a freight of concepts
entirely new to me. This book could not
have existed without High Ground.

Shaping Things was particularly shaped
by a High Ground conversation with
Scott Klinker of Cranbrook, whose disci-
plined thinking was of great help in
showing me where my inchoate notions
ought to go.

After many years of admiration from
afar, I would like to thank Brenda Laurel
for her kindness, hospitality and impressive
energy during my residency at Art Center
College of Design. All my colleagues at this

fine institution have been remarkably supportive of their school's first "Visionary in Residence", but Brenda Laurel is the visionary's visionary.

Peter Lunenfeld commissioned this book as part of his admirable series and was actively involved in practically every aspect of every page. It was delightful to work with Lorraine Wild after many years of admiring her comprehensive grasp of graphic design.

I owe a debt to my many online friends and allies in "Viridiandesign.org" and the "Viridian Curia"—especially the hard-charging Bright Green activists of "Worldchanging.com," including Jamais Cascio, Alex Steffen, Jon Lebkowsky and Dawn Danby.

If one somehow feels a need to become a design theorist, it really helps to find some people who (a) take the ideas seriously and (b) can put those ideas into direct practice. Few things in my life have given me as much satisfaction as the advent of "Worldchanging."

I have learned a lot in a hurry from Mimi Ito and her Southern California Digital Culture Group. Beneath the phony tinsel of Los Angeles lurks the genuine tinsel of Los Angeles— a glittering fiber-optic network with impressively high bandwidth.

It may sound a little odd to thank publications and institutions such as *Wired* magazine, *Metropolis*, the IDSA, ACM SIGGRAPH, and the AIGA, but behind those mastheads and acronyms are a host of real people who allowed me write articles and pound podiums. Without that tangled nexus of technosocial support, I would never have been able to write a book like this.

Adele Winstel always wondered about my odd authorly habits in dedicating books. Eventually books of her own appeared. We never own the torch, Adele, we can only pass the fire. Brilliancy, speed, lightness and glory, Adele: this book is for you.

Bruce Sterling

Colophon

Shaping Things was designed by Lorraine Wild
and Stuart Smith of Green Dragon Office,
Los Angeles during the Spring of 2005. The
typefaces used throughout this book are
Apex Sans, Apex Serif (designed by chester and
Rick Valicenti of Thirst, Chicago, Illinois, 2003),
and BALLON DROP SHADOW (Max R. Kaufman for
the American Type Foundry, 1939), Dom Casual
(designed by Pete Dom, 1951–53), Keedy Sans
(designed by Mr. Keedy of Los Angeles,
California, published by Émigré, 1989), LO-RES
(Zuzana Licko of Émigré, Berkeley, California,
1985 [revised 2001]), PLAYBILL (after Robert
Harling, 1938), School Script (anonymous
freeware) and VAG Rounded (Adrian Williams,
1979) and was typeset on a 1.8 GHz Apple Power
Mac G5 with a 17 inch Apple LCD Digital Studio
Monitor using Adobe InDesign CS software.
The cover photograph was taken using a
Canon Digital Rebel 6.3 mega-pixel SLR and
manipulated with Adobe Photoshop CS
(the preliminary photographs "captured" for the
cover of this book, using a PalmOne Treo 600
smart phone [not the 650 , dammit] were rejected
by the editorial director for being "too low-res.").
The charts in this book, based on original ink
and graph paper drawings of Bruce Sterling,
were re-interpreted by Hilary Greenbaum of
Green Dragon Office, using Adobe Illustrator CS.
This book was printed by Quebecor World on
80 lb. Jenson Gloss White (interior) and 10 pt
Cornwall C1S (cover) under the supervision of
The MIT Press.

© 2005
Massachusetts Institute of Technology

MIT Press books may be purchased at special
quantity discounts for business or sales
promotional use.

For information, please email
<special_sales@mitpress.mit.edu>

or write to
Special Sales Department,
The MIT Press, 55 Hayward Street,
Cambridge, MA 02142.

was printed and bound in the United States.

Library of Congress
Cataloging-in-Publication data

Sterling, Bruce
  Shaping Things/Bruce Sterling
    p. cm
  ISBN-13 978-0-262-19533-1 (hc.: alk. paper)
  978-0-262-69326-4 (pbk.: alk. paper)
  ISBN 0-262-19533-X (hc.: alk. paper)—
  ISBN 0-262-69326-7 (pbk.: alk. paper)
    1. Technological forecasting 2. Technology
  assessment I. Title.

T174.5772005
745.2—dc22   200505109

10 9 8 7 6 5 4